“Solution Of The Riemann Hypothesis And Prime Numbers”

José Mauricio Orellana Díaz

However, as it is written: "What no eye has seen, what no ear has heard, and what no human mind has conceived" the things GOD has prepared for those who love him. 1Corinthians. 2:9 NIV.

Contents

0.1 Acknowledgements

First of all, I thank God for giving me the wisdom, understanding, and intelligence to develop a new methodology to analyze differently one of the problems of the millennium, such as The Riemann Hypothesis and Prime Numbers. In his Word, we learn that the fear of the Lord is the beginning of knowledge, but fools despise wisdom and instruction. Proverbs 1:7. On several occasions I was confused and in chaos for giving a solution in a different way than The Riemann Hypothesis and Prime Numbers. In my prayers, I asked God to give me wisdom, and by His grace I received it. If any of you lacks wisdom, you should ask God, who gives generously to all without finding fault, and it will be given to you. James 1:5. Solomon's Proverbs teach us: Get wisdom, get understanding; do not forget my words or turn away from them. Proverbs 4:5. Thank God for His Word that tells us so: Instruct the wise and they will be wiser still; teach the righteous and they will add to their learning. Proverbs 9:9. For the wisdom of this world is foolishness in God's sight. As it is written: "He catches the wise in their craftiness" 1 Corinthians 3:19. How much better to get wisdom than gold, to get insight rather than silver!. Proverbs 16: 16. The mind of a discerning person gains knowledge, while the ears of wise people seek out knowledge Proverbs 18:15. From the depths of my being I thank God for every blessing, knowledge, wisdom and intelligence that He has given me; to understand many physical and mathematical topics. It is God who alters the times and seasons, and he removes kings and promotes kings. He gives wisdom to the wise and knowledge to the discerning. Daniel 2:21.

0.2 Dedications

As a preview of my second mathematical work, thanks to the knowledge and wisdom that God has given me. I dedicate the second mathematical work to my Lord and God because it has allowed me to write this second work, even when sometimes I was confused and in chaos trying to give a solution to a problem as complex as the Riemann Hypothesis and Prime Numbers. I spent a lot of time writing in my notebooks and making many corrections in those moments of difficulty, confusion, and chaos regarding such a complex mathematical problem. Today, I have achieved the goal, thanks to your kindness and favors, I can thank you infinitely, for giving me the capacity and allowing me to write these interesting theories of Trigonometric Spirals, their Angles, and Trigonometric Partitions that can be applied to The Riemann Hypothesis and Prime Numbers. I also dedicate this second work to my children, whom I will always love and who are always in my thoughts and my soul, which are my special treasures and inheritance that my God has given me. Hoping that you can read it and learn from the creativity and talents that God has given me in the field of mathematics. God grants wisdom to my special treasures and has the prize of being wise, intelligent, and understanding in this world , making good use of your time each day. To God be the glory for what He did in me, thank you for His love and infinite mercy, with my life, since it has been Him, who has helped me reach this final goal of finishing this book, with these new and interesting theories for the mathematical physics field.

0.3 Summary

The equations of the Theory of Spiral Angles, Spirals, and Trigonometric Partitions make it possible to analyze the Prime Numbers and the Riemann Hypothesis in a different way. By applying these equations, we can define, in the plane of real and imaginary numbers, new formulas that help us analyze the behavior of the functions of the Prime Numbers and the Riemann Zeta Function. In this study developed for both topics, a strong relationship will be observed between the Prime Numbers and the Riemann Hypothesis. The Theory of Spiral Angles, Spirals, and Trigonometric Partitions are seven mathematical equations that have different applications depending on the case study. In this study, two of the seven equations with which we analyze the Prime Numbers and the Riemann Z Function are used. These equations are capable of studying equations that are periodic and non-periodic without limits of periods, frequency, angular velocity, and time. The first is the Radius Growth Partitions equation which is applied to define an equation for Prime Numbers. This radius growth partitions equation counts its initial radius, final radius, spiral angle, and trigonometric partitions. This is an analysis of various flat circular waves; which are undergoing diametric cuts or partitions, generating a co-secant line. With the spiral angles, various equations can be obtained that relates to the trigonometric partitions, or the posterior prime number minus the previous prime number, and the exponent of the prime numbers. The second is The Chord Partitions which is applied to define an equation for the Riemann Zeta Function. With both equations of the trigonometric partitions, we can analyze the behavior of the Prime Numbers and the Riemann Zeta Function, both in the plane of real numbers and in the plane of complex numbers. The equation of the Trigonometric Partitions of the Chord, includes the final radius, initial radius, spiral angle, and trigonometric partitions. This is an analysis of various plane circular waves; which are undergoing diametric cuts or partitions generating a co-secant line. Therefore, with this new methodology and these new mathematical theories; trivial zeros can be studied, as well as non-trivial ones of the equations both in the plane of real numbers and in the plane of complex numbers.

0.4 Introduction

The equations of the Theory of Spiral Angles, Spirals, and Trigonometric Partitions make it possible to analyze the Prime Numbers and the Riemann Hypothesis in a different way. By applying these equations, we can define, in the plane of real and imaginary numbers, new formulas that help us analyze the behavior of the functions of the Prime Numbers and the Riemann Zeta Function. In this study developed for both topics, a strong relationship will be observed between the Prime Numbers and the Riemann Hypothesis, which is the Module of Length of the Prime Numbers, equivalent to the difference of the largest (posterior) prime number minus the smallest (previous) prime number, related to the growth of the radius and the trigonometric partitions of the chord. Therefore, in this study the different equations that work as flat circular waves of this new theory can be used. On the plane of real numbers, the initial angles selected for these equations can be fixed and transformed in their initial analyses. Likewise, you can use equations with variable initial angles. It is important to note that these equations in the real number plane have a different behavior than the complex number plane. Consequently, in mathematical functions, their variables can be changed, since some of them will have their graphs visible in the imaginary plane and other functions will be invisible in the same plane, calling this concept "transformation of the mathematical function ." For the Riemann Zeta Function, four equations generate the graph of the Riemann Zeta function, in the plane of imaginary numbers, amazing things happen, such as, for example, when we analyze at 90° degrees, this equation forms a graph with a single formula, this is not the case for 60° degrees, where two equations are required, one positive and one negative, to generate the same graph. Finally, a general formula for the Riemann Zeta Function is obtained, to provide an innovative solution to a millennium problem.

Chapter 1

Prime Numbers Equation

1.1 Theory Of Spiral Angles, Spirals, And Trigonometric Partitions

Some numbers have the special property that they cannot be expressed as the product of two smaller numbers, e.g., 2, 3, 5, 7, etc. Such numbers are called prime numbers, and they play an important role, both in pure mathematics and its applications. The distribution of such prime numbers among all natural numbers does not follow any regular pattern. However, the German mathematician G.F.B. Riemann (1826 - 1866) observed that the frequency of prime numbers is very closely related to the behavior of an elaborate function $\zeta(s) = 1 + \frac{1}{2^s} + \frac{1}{3^s} + \frac{1}{4^s} + ...$ Called the Riemann Zeta function. The Riemann hypothesis asserts that all interesting solutions of the equation $\zeta(s) = 0$ lie on a certain vertical straight line. This has been checked for the first 10,000,000,000,000 solutions. A proof that it is true for every interesting solution would shed light on many of the mysteries surrounding the distribution of prime numbers. The Clay Mathematics Institute, March 31st, 2023, Riemann Hypothesis,(Clay Mathematics Institute, n.d.-a)

https://www.claymath.org/millennium-problems/riemann-hypothesis.

It's the Theory of Spiral Angles, Spirals, and Trigonometric Partitions; which can study and analyze both periodic and non-periodic functions, without limits on periods, frequency, angular velocity, and time, therefore, it was developed in a circular form; similar to the waves generated in water by a potential energy. Seven mathematical equations help us to comprehensively study the components of a circle, which is continuously divided in to halves indefinitely. The part that continuously divides the circle to study it insections is called Trigonometric Partitions, which can connect all planes into one; such as the planes of real and imaginary numbers. Therefore, for more knowledge of these theories, one should read the book I have written at Barkerand Jules. I hold the copyright for it.Library Of Congress Control Number : 1-12354195611 in the USA.(OrellanaDíaz,2023)

If the radius is inversely proportional to the spiral angle or equivalent to the trigonometric function of the secant, this equation in the polar form will be able to generate a trigonometric spiral Therefore;$r(\varphi) = 1/\cos(\varphi) = \sec(\varphi)$. Where its integrating factor is $\int_0^{\varphi} d\varphi / r(\varphi)$;replacing $r(\varphi)$: $r = r_0 * e^{\int_0^{\varphi} \cos(\varphi) * d\varphi}$ this is a definite integral with $\varphi_0 = 0$, in r_0, and with $r = \sqrt{(m^2 + n^{2s})}$.

If we integrate the integrating factor we will obtain:$r = r_0 * e^{\sin(\varphi)}$, solving for $r_0 = r * e^{-\sin(\varphi)}$,Substituting r we will obtain:$r_0 = \sqrt{m^2 + n^{2s}} * e^{-\sin(\varphi)}$, Converting r_0 into a function, therefore it will be:$r(s) = \sqrt{m^2 + n^{2s}} * e^{-\sin(\varphi)} *$, with **k** turns, its polar coordinates; Similarly, in these equations $\varphi = \omega * t = 2 * \pi * f * t = k * x - \omega * t = 2 * \pi * n$ with $k = 2 * \pi / \lambda$ can be used. When **k** and φ increase, its revolution and its frequency increase, and the intensity of the black color in the graphs of the 3D planes will also increase.

$r(s) = \sqrt{m^2 + n^{2s}} * e^{-\sin(\varphi)}$. The green color graph in GeoGebra.

$r_x(s) = \cos(k * \varphi) * \sqrt{(m^2 + n^{2*s})} * e^{-\sin(k*\varphi)}$. The purple color graph in GeoGebra.

$r_y(s) = \sin(k * \varphi) * \sqrt{(m^2 + n^{2*s})} * e^{-\sin(k*\varphi)}$. The blue color graph in GeoGebra.

These three equations in the 2D plane will start as waves. Except for its polar coordinate, which will be a linear radial equation (red color in GeoGebra); Later when **n** takes negative numbers it will become a point (red) or particle and the waves disappear. In the 3D plane in the polar equation with negative **n** values, the solid of revolution will become its X-ray of the solid of revolution, which is equivalent to X-rays the solid of revolution in the 3D plane. This equation is very consistent even if raised to its power s and renamed R(s).

$R(s) = r^s(s) = \sqrt{(m^2 + n^{2*s})^s} * e^{-s*\sin(k*\varphi)}$,

$R_x(s) = \cos(k*\varphi) * \sqrt{(m^2 + n^{2*s})^s} * e^{-s*\sin(k*\varphi)}$,

$R_y(s) = \sin(k*\varphi) * \sqrt{(m^2 + n^{2*s})^s} * e^{-s*\sin(k*\varphi)}$.

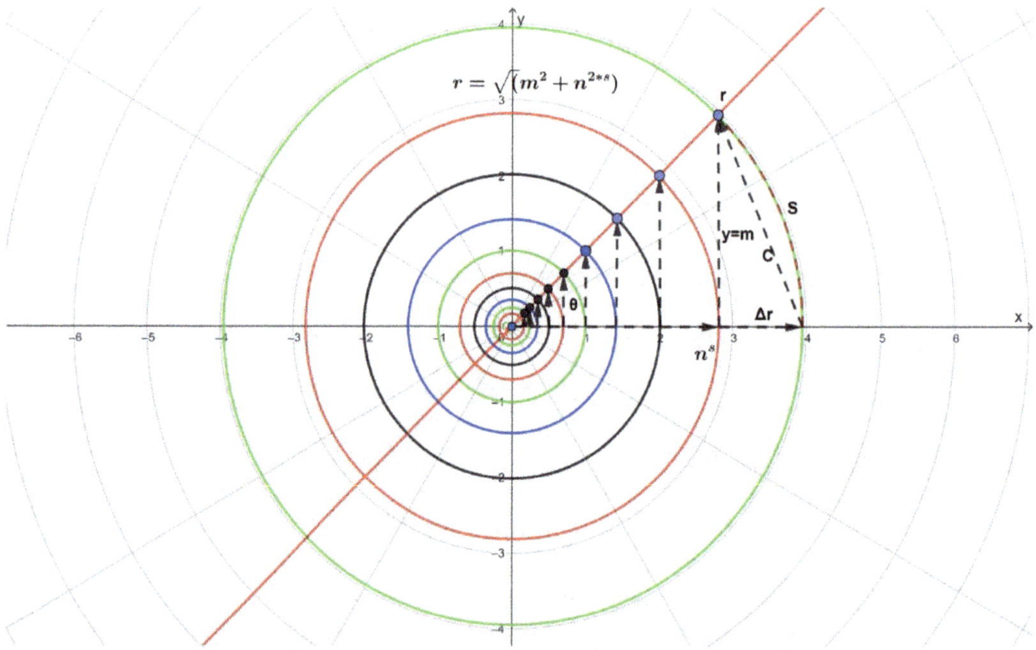

Figure 1.1: Circular wave graph

All basic trigonometric identities can be transformed into trigonometric identities in terms of the trigonometric partitions p.

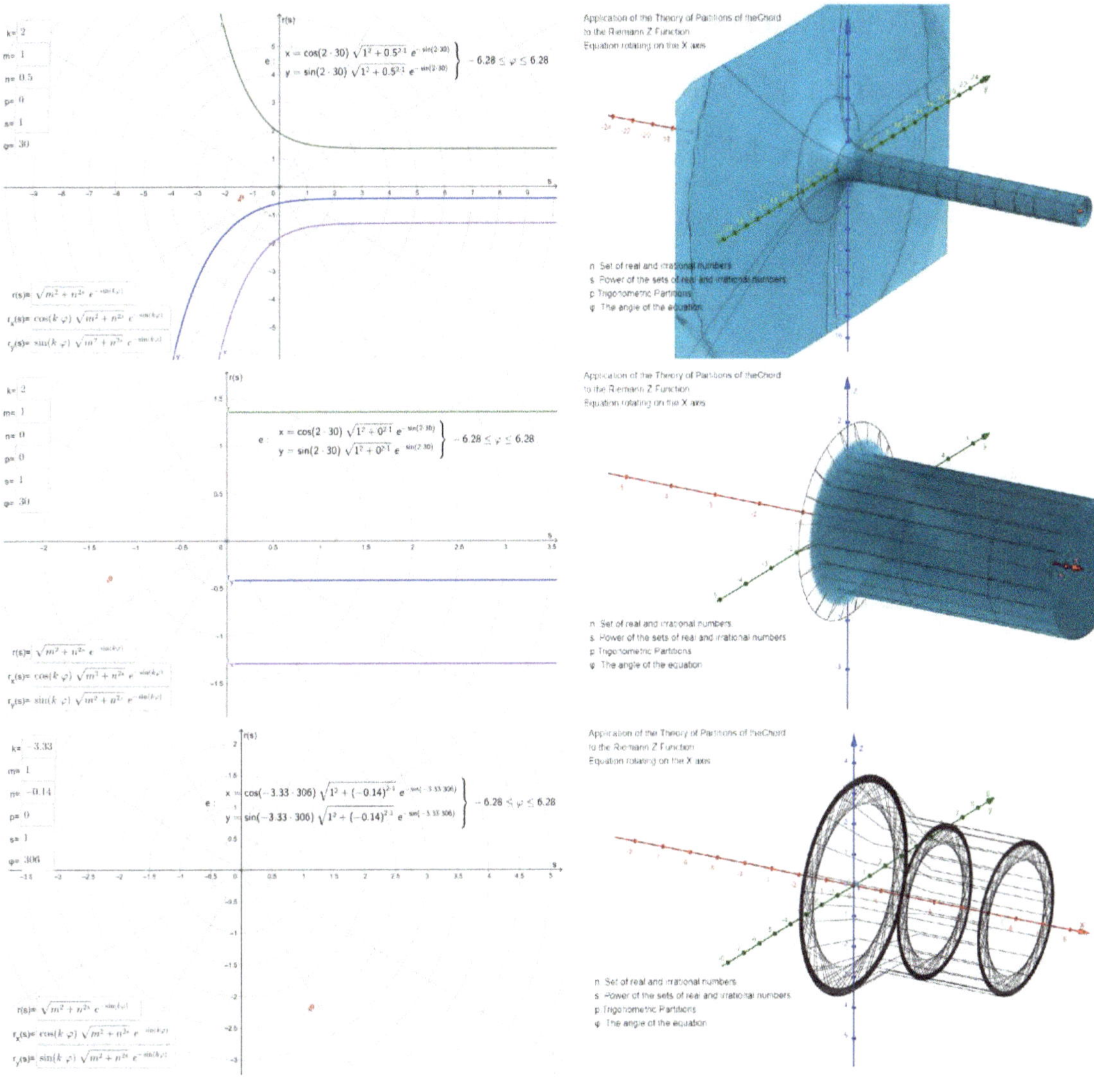

Figure 1.2: Trigonometric Partitions Graphics.

Theory Of Trigonometric Partitions

Trigonometric Identities Of Mid-Angles:

The unit circle is continuously divided into several halves, yielding new trigonometric identities for mean angles. With this methodology of continuously dividing the circle, we can develop a new theory to analyze the Cartesian plane with trigonometric partitions. With this theory of trigonometric partitions, greater precision can be given to the equation, yielding exact results. Similarly, it is possible to study the pattern of a mathematical function, search for the solution of a difficult mathematical equation using trial and error methodology, and finally, develop a simpler and more understandable mathematical equation for the wavelet transform. If we continuously make radial cuts on a circle, we can divide it into successive halves. This physical effect of continuously cutting the circle is called "Trigonometric Partitions." This physical effect of continuous cuts can be executed radially for the apothem, arrow, and radius growth. Another way to make cuts is in the direction of the chord, circular sector, and along the X and Y axes. The cut of the chord is called "Chord Trigonometric Partition," the cut of the circular sector is called "Circular Sector Trigonometric Partition," the cut of the arrow is called "Arrow Trigonometric Partition," the cut of the apothem is called "Apothem Trigonometric Partition," and so on. This physical effect of trigonometric cuts will be denoted by the letter "p." The physical effect of "p" can be considered positive or negative depending on the direction of the cut or rotation of its angular velocity. If we make radial cuts of the circle, when "p" equals zero, then its physical form is a circle; if "p" equals one, then the circle has been divided into two halves; if "p" equals two, the circle has been divided into four equal parts, and so on. This is the general trigonometric identity for trigonometric partitions (p). Where p represents the physical effect of dividing or cutting a circle into infinite parts.

With $p = 0$, $2 * \sin(\theta/2) = \sqrt{2} * \sqrt{1 - \cos(\theta)}$ and $2 * \sin(\theta/2^{p+1}) = \sqrt{2} * \sqrt{1 - \cos(\theta/2^{p})}$.

Where $2 * \sin^2(\theta/2^{p+1}) = 1 - \cos(\theta/2^{p})$, Therefore,$\cos(\theta/2^{p}) = 1 - 2 * \sin^2(\theta/2^{p+1})$.

Defining p as the trigonometric partitions in a circle; p can be any number. The following general trigonometric identity for mid-angles can be demonstrated with the following graph, starting with p=0 then; $2 * \sin(\theta/2^{p+1}) = \sqrt{2} * \sqrt{1 - \cos(\theta/2^p)}$, solving for the cosine of the angle we can obtain: $\cos(\theta/2^p) = 1 - 2 * \sin^2(\theta/2^{p+1})$ or $\sin(\theta/2^{p+1}) = \sqrt{2} * \sqrt{1 - \cos(\theta/2^p)}/2$. Similarly, this concept applies to the trigonometric identity of the unit circle $\cos^2(\theta/2^{p+1}) + sin^2(\theta/2^{p+1}) = 1$. If we substitute $\sin^2(\theta/2^{p+1}) = [1 - \cos(\theta/2^p)]/2$ into the general trigonometric identity for the unit circle, we will get $\cos(\theta/2^p) = 2 * \cos^2(\theta/2^{p+1}) - 1$. For example, as a reference point for the trigonometric partitions of the circular sector, the chord, and the radius growth, we take the initial angle Theta $\theta_0 = 0$. Also, the final angle Theta can be considered as $\theta = 2 * \pi$, in case more than 360° degrees or several turns are desired, the letter n will be used to define the equation of the angle Theta as $\theta = 2 * \pi * n$. These trigonometric partition equations of the angle can be used about the number of turns, angular velocity, frequency, time, position x, wavelength, and number of waves $\theta = 2 * \pi * n = \omega * t = 2 * \pi * f * t = k * x - \omega * t$ with $k = 2 * \pi/\lambda$. Rewriting the formulas, depending on the study being conducted, for example; using with its number of turns n: If we use $\theta = 2 * \pi * n$, in the following equation $\cos(\theta/2^p) = 1 - 2 * \sin^2(\theta/2^{p+1})$ we will obtain $cos(\pi * n/2^{p-1}) = 1 - 2 * \sin^2(\pi * n/2^p)$, in $\cos(\theta/2^p) = 2 * \cos^2(\theta/2^{p+1}) - 1$ we will obtain $\cos(\pi * n/2^{p-1}) = 2 * \cos^2(\pi * n/2^{p-1})$. Therefore, we can do the same with the angular velocity and time $\theta = \omega * t$ substituting in $\cos(\theta/2^p) = 1 - 2 * \sin^2(\theta/2^{p+1})$ we will obtain $\cos(\omega * t/2^p) = 1 - 2 * \sin^2(\omega * t/2^{p+1})$ or $\cos(\omega * t/2^p) = 2 * \cos^2(\omega * t/2^{p+1}) - 1$.

Circular diagram to develop general mid-angle trigonometric identities.

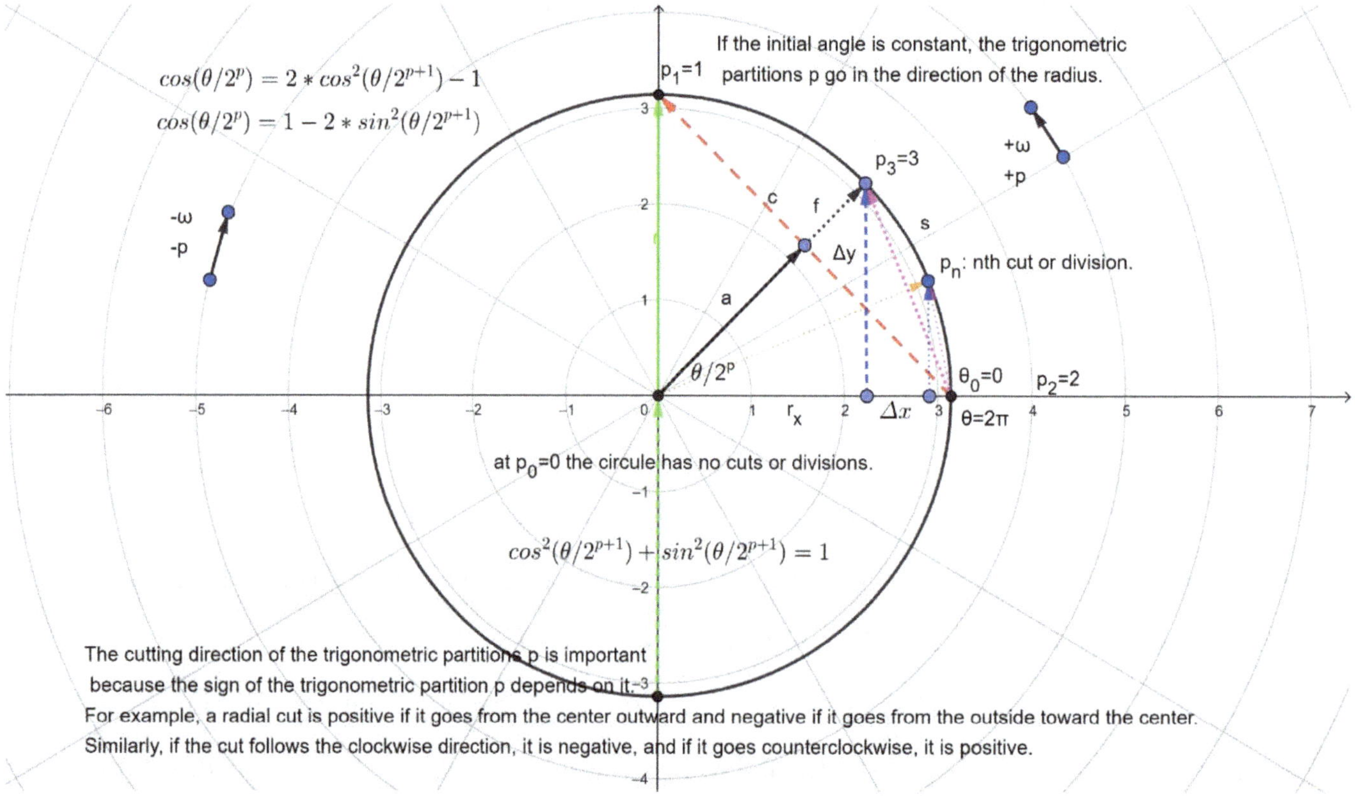

Figure 1.3: With this graph, all equations for mid-angles in a general form can be developed and the type of trigonometric cuts that can be made can be understood.

This advanced mathematical concept can be applied in classical and quantum physics to analyze all equations containing a radius in their formulas. Similarly, we can do the same with frequency and time $\theta = 2*\pi*f*t$ by substituting into $\cos(\theta/2^p) = 1-2*\sin^2(\theta/2^{p+1})$ to obtain $\cos(\pi*f*t/2^{p-1}) = 1-2*\sin^2(\pi*f*t/2^p)$ or $\cos(\pi*f*t/2^{p-1}) = 2*\cos^2(\pi*f*t/2^p)-1$. Finally, we can do this with the number and wavelength in position x as a function of time $\theta = k*x-\omega*t$ with $k = 2*\pi/\lambda$ by substituting into $\cos(\theta/2^p) = 1-2*\sin^2(\theta/2^{p+1})$ to obtain $\cos[(k*x-\omega*t)/2^p] = 1-2*\sin^2[(k*x-\omega*t)/2^{p+1}]$ or $\cos[(k*x-\omega*t)/2^p] = 2*\cos^2[(k*x-\omega*t)/2^{p+1}]-1$. Lastly, we have three general trigonometric identities for mid-angles in terms of trigonometric partitions, which are: $2*\cos(\theta/2^{p+1}) = \sqrt{2}*\sqrt{[1+\cos(\theta/2^p)]}$,

$2 * \sin(\theta/2^{p+1}) = \sqrt{2} * \sqrt{[1 - \cos(\theta/2^p)]}$,

$\cos^2(\theta/2^{p+1}) + \sin^2(\theta/2^{p+1}) = 1$,

Regarding these equations, if we change the base (b) 2 to base (b) 3, this definition still holds for double trigonometric partitions, but not for triple trigonometric partitions:

$\cos^2(\theta/3^{p+1}) + \sin^2(\theta/3^{p+1}) = 1$,

$\cos^2(\theta/b^p) + \sin^2(\theta/b^p) = 1$,

$\cos^2(\theta/2^p) + \sin^2(\theta/2^p) = 1$,

If the trigonometric identities hold for any angle value and partitions.

$2 * \cos^2(\theta/b^{p+1}) = 1 + \cos(\theta/b^p)$.

$2 * \sin^2(\theta/b^{p+1}) = 1 - \cos(\theta/b^p)$.

Table 1.1: Proving Trigonometric Identities With Values.

θ=	35	
b=	2	
p=	0	Sums
$\cos^2(\theta/b^p)$	0.671010072	1.000000000
$\sin^2(\theta/b^p)$	0.328989928	
$\cos^2(\theta/b^{p+1})$	0.909576022	1.000000000
$\sin^2(\theta/b^{p+1})$	0.090423978	
$2 * \cos^2(\theta/b^{p+1})$	1.819152044	
$1 + \cos(\theta/b^p)$	*1.819152044*	
$2 * \sin^2(\theta/b^{p+1})$	0.180847956	
$1 - \cos(\theta/b^p)$	*0.180847956*	

$$\cos^2(\theta/b^{p+1}) + \sin^2(\theta/b^{p+1}) = 1,$$

$\cos^2(\theta/3^{p+1}) + \sin^2(\theta/3^{p+1}) = 1,$

The trigonometric identities are not fulfilled.

$2 * \cos^2(\theta/b^{p+1}) \neq 1 + \cos(\theta/b^p).$

$2 * \sin^2(\theta/b^{p+1}) \neq 1 - \cos(\theta/b^p).$

Table 1.2: Proving Trigonometric Identities With Values.

θ=	35	
b=	3	
p=	0	Sums
$\cos^2(\theta/b^p)$	0.671010072	1.000000000
$\sin^2(\theta/b^p)$	0.328989928	
$\cos^2(\theta/b^{p+1})$	0.959108053	1.000000000
$\sin^2(\theta/b^{p+1})$	0.040891947	
$2 * \cos^2(\theta/b^{p+1})$	1.918216107	
$1 + \cos(\theta/b^p)$	*1.819152044*	
$2 * \sin^2(\theta/b^{p+1})$	0.081783893	
$1 - \cos(\theta/b^p)$	*0.180847956*	

Wavelets and Multiple-wavelet Transforms.

Multiple Wavelet Transform:

The graph in green represents the trigonometric partition graph of the radius as a function of angular velocity. The graph in red represents the X component graph of the trigonometric partitions of the radius as a function of angular velocity. The graph in blue represents the Y component graph of the trigonometric partitions of the radius as a function of angular velocity. In a video on social media, I demonstrate how the theory of chord trigonometric partitions can generate a wide variety of wavelets, both single wavelets and multiple wavelets. This is the easiest tool to generate wavelets and analyze each of the variables and their effects on the graph. The same can be done with other equations of trigonometric partitions of the circular segment, the arrow, the apothem, and the radius growth..., thus causing the family of wavelets to expand. One of the characteristics of this new methodology is that we can easily generate polar functions and analyze these mathematical equations of trigonometric partitions of the chord in the third dimension. This methodology offers us the ease of analyzing each of the nine variables in this equation, where:

k: Torsions or twists.

n: The number of waves, this is one since I have several equations written in GeoGebra, which is why it doesn't animate automatically, as it does not cause any effect because this number of waves is a function of the wavelength of the wavelet transform. $n = 2 * \pi / \lambda$.

x_0: The position of the wavelet on the X-axis.

ω: Angular velocity, this is one since I have several equations written in GeoGebra, which is why it doesn't animate automatically, as it does not cause any effect because these trigonometric partition equations are of the radius as a function of angular velocity $r(\omega)$.

t: Time.

p: Trigonometric partitions of the chord.

r_0: The amplitude or initial radius. This can also be applied as the final radius in decrease or compression.

g: The position of the multiple or transformed wavelets on the Y-axis.

λ: Wavelength of the wavelet transform.

Another advantage of this system is that with this equation of trigonometric partitions, we can use all the equivalent angle equations $\theta = \omega * t = 2 * \pi * f * t = n * x - \omega * t...$, and we will always be able to generate wavelets. We can even place the equations of the radius partitions as a function of most of the variables of the formula to generate their wavelets. With this trigonometric partition methodology, we will be able to generate the uncertainty principle graph, double slit experiment graph, applying this theory to classical and quantum physics by analyzing all those mathematical equations that have radii in their equations. This new methodology does not require complex numbers since the equation of the radius of the trigonometric partition of the Y component comes to replace complex numbers. Similarly, the Fourier series, or its transforms are not required since these equations were developed as concentric circles like the waves generated by a stone falling into the water, carrying the physical effect of cuts in every sense of a component of a circle as a function of the radii of the trigonometric partitions. In this particular case, if we derive the equation of the trigonometric partitions of the chord of the radius as a function of the angle, we will obtain the chord as a function of the trigonometric partitions. This methodology helps us compare all the equations of the radius of the trigonometric partitions based on each of the variables that can be formed, for example, the radius as a function of velocity $r(\omega)$, the radius as a function of time r(t)... and generate a wide variety of wavelets with their characteristics. The same happens with the equation of the trigonometric partitions of the chord as a function of frequency and time; r(f), r(t). In these equations of this new methodology, it is not required that the functions be periodic as in the case of Fourier series. The functions can be analyzed periodically and non-periodically, without limits on the domain of their values for each variable used in these equations of trigonometric partitions of the chord. The components of the radius as a function of the X coordinate and the Y coordinate are perpendicular, therefore they are orthogonal, and they allow arithmetic properties of addition, subtraction, multiplication, and division for the equations of the X and Y components of the radius of the trigonometric partitions, fulfilling all the conditions of a wavelet. Graphs of the equations of the trigonometric partitions of the Chord as a function of frequency and time, all of these graphs are made with pure mathematical equations in GeoGebra:

$r(x_0) = r_0 * e^{[-2^{p+2}*\sin^2((2*\pi*x_0/\lambda-\omega*t)/2^{p+1})]} + g,$

$r_x(x_0) = r_0 * \cos[k * (2 * \pi * x_0/\lambda - \omega * t)] * e^{[-2^{p+2}*\sin^2((2*\pi*x_0/\lambda-\omega*t)/2^{p+1})]} + g,$

$r_y(x_0) = r_0 * \sin[k * (2 * \pi * x_0/\lambda - \omega * t)] * e^{[-2^{p+2}*\sin^2((2*\pi*x_0/\lambda-\omega*t)/2^{p+1})]} + g,$

Parametric curve:

$a = Curve(r_x(x_0), r_y(x_0), x_0, -2 * \pi, 2 * \pi),$

$r(t) = r_0 * e^{[-2^{p+2}*\sin^2((2*\pi*x_0/\lambda-\omega*t)/2^{p+1})]} + g,$

$r_x(t) = r_0 * \cos[k * (2 * \pi * x_0/\lambda - \omega * t)] * e^{[-2^{p+2}*\sin^2((2*\pi*x_0/\lambda-\omega*t)/2^{p+1})]} + g,$

$r_y(t) = r_0 * \sin[k * (2 * \pi * x_0/\lambda - \omega * t)] * e^{[-2^{p+2}*\sin^2((2*\pi*x_0/\lambda-\omega*t)/2^{p+1})]} + g,$

Parametric curve:

$a = Curve(r_x(t), r_y(t), t, -2 * \pi, 2 * \pi),$

$$r(\omega) = r_0 * e^{[-2^{p+2}*\sin^2((2*\pi*x_0/\lambda-\omega*t)/2^{p+1})]} + g,$$

$r_x(\omega) = r_0 * \cos[k * (2 * \pi * x_0/\lambda - \omega * t)] * e^{[-2^{p+2}*\sin^2((2*\pi*x_0/\lambda-\omega*t)/2^{p+1})]} + g,$

$r_y(\omega) = r_0 * \sin[k * (2 * \pi * x_0/\lambda - \omega * t)] * e^{[-2^{p+2}*\sin^2((2*\pi*x_0/\lambda-\omega*t)/2^{p+1})]} + g,$

Parametric curve:

$a = Curve(r_x(\omega), r_y(\omega), \omega, -2 * \pi, 2 * \pi).$

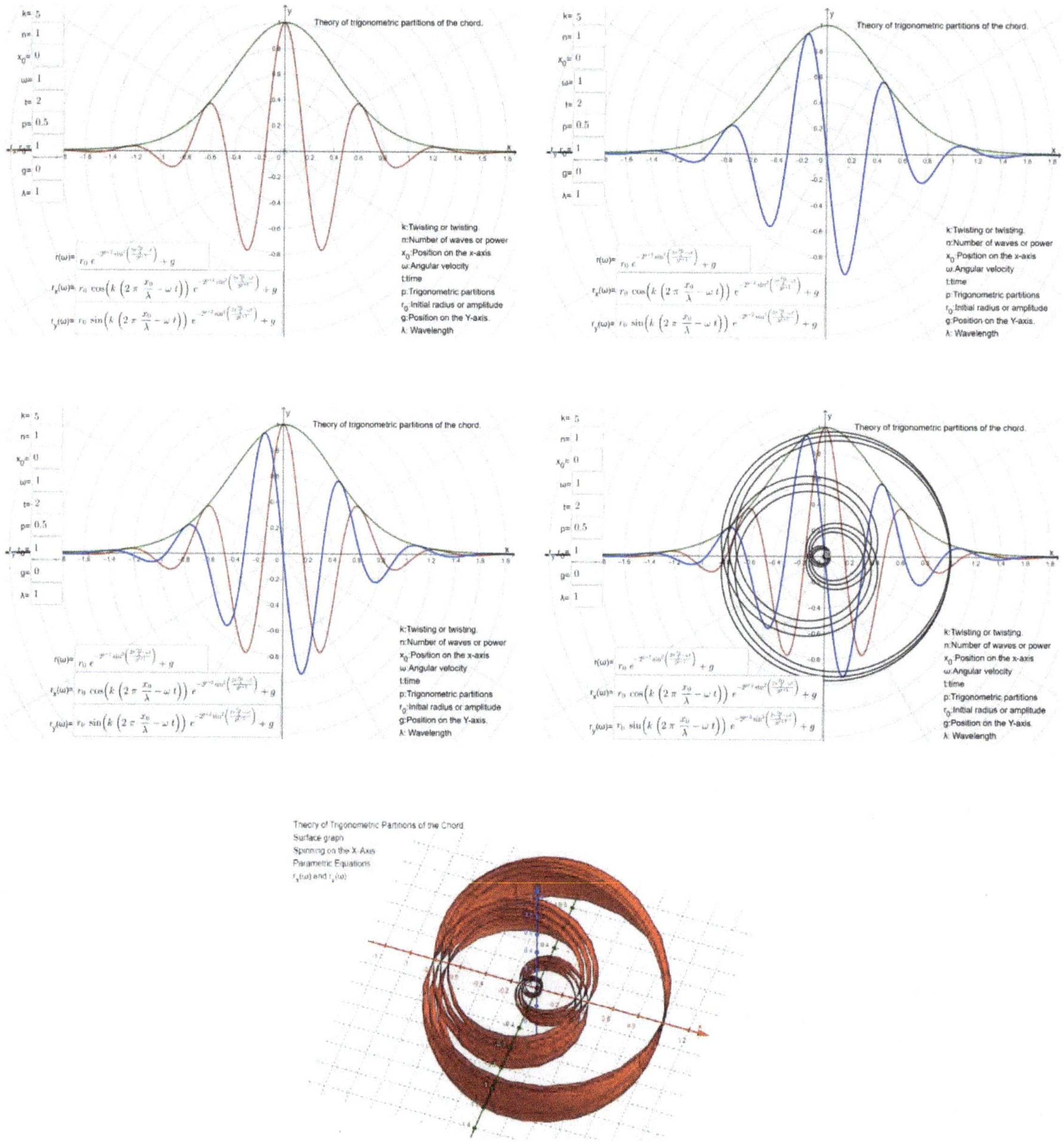

Figure 1.4: Wavelet Charts with Trigonometric Partitions.

Wavelets as a function of Frequency, Angular Velocity, and Time.

Wavelet Transforms and Multiwavelet:

If we ask ourselves: How are wavelets generated? These equations of trigonometric chord partitions come from concentric circular analysis, similar to the waves generated in water when a stone falls. With them, we can study the decrease of the final radius, or rearrange the original formula of the chord partition to study the initial radius and analyze its birth or the growth of the initial radius, which is why the exponent 2 is negative. With the Theory of Trigonometric Partitions, we can easily generate and analyze the wavelets and their transforms.There are several equations of trigonometric partitions; in this particular case, the Trigonometric Chord Partition is being used. With this new concept, we can perform 2D analysis with its parametric curves of one equation for the X coordinate and another for the Y coordinate, as well as in 3D with the technique of solid revolution which is equivalent to performing an integral. In these equations of the radius of trigonometric partitions, we will use the equivalent equations of the angle as a function of frequency and time to generate multiple wavelets, likewise with the equivalent equation of the angle as a function of the constant k; which in this case is equal to the rotations, twist, and wave numbers, the position x, of the Angular Velocity ω, and time t; various wavelets can be generated. By studying each of the characteristics, for example, if we increase the frequency, we observe how the wavelet compresses, observing the effects that each variable causes in the mathematical equations. In these equations of trigonometric partitions of the wavelets, we can write the radius as a function of frequency, and the radius as a function of time, and we will always generate wavelets with their characteristics. Therefore, if we write these equations of the radius as a function of trigonometric partitions p, and the constant k, wavelet transforms are not generated. Similarly, in the equation as a function of position and time, we can use all the variables to write the radius as a function of each of them, and we will always generate wavelets. With the first derivative of the trigonometric partitions of the chord as a function of the trigonometric partitions p, we can generate scalograms. This same new methodology can be used in the equations of trigonometric partitions of the circular sector, apothem, arrow, and radius growth, creating a wide variety of wavelets and their transforms. This is a new technique that does not require software for its programming, complex numbers, or analysis with the Fourier series or its transforms. The tool used is GeoGebra, initially, I started making them in Excel spreadsheet. With these same equations, we can create the graph of the uncertainty principle; in the graphs generated by this theory, you will observe the distortion caused by the harmonics in the sine wave, and the DC offset of the equation of the Y coordinate. This knowledge can be applied to vibration analysis, Goldbach's conjecture etc. If you want to learn more about the Theory of Trigonometric Partitions, you should read the book I have written on this and other topics of its application. Wavelets as a function of frequency, angular velocity, and time. With these equations of trigonometric chord partitions of the radius as a function of time, it can be compared with the Fast Fourier Transform (FFT). The equation of trigonometric chord partitions of the radius as a function of angular velocity can be compared with the Discrete Fourier Transform (DWT). With these new equations of trigonometric partitions, it will be possible to study quantum entanglement and quantum tunneling. With this new technique, we can use integers for the variables of the equation and create wavelets. Even if we increase the power of this equation of trigonometric partitions, wavelets and the graph of the uncertainty principle can continue to be generated. This is pure mathematics, and with these equations of trigonometric partitions, we can apply arithmetic properties and create wavelets. The GeoGebra sliders are the domain of each variable, and in these equations, they have no limits. Any value from negative infinity to positive infinity can be used. With this methodology, wavelets generated at low and high frequencies can be analyzed. The Theory of Trigonometric Partitions is a super-symmetric equation; it can behave like a dot, a straight line, or a curve, and works in the real plane as well as in the complex number plane. Graphs of the equations of the trigonometric partitions of the Chord as a function of frequency and time, all of these graphs are made with pure mathematical equations in GeoGebra:

$R_2(t) = r_0^n * e^{[-n*2^{p+2}*\sin^2(\pi*f*t/2^p)]}$,

$R_{2x}(t) = r_0^n * e^{[-n*2^{p+2}*\sin^2(\pi*f*t/2^p)]} * [cos(2*\pi*k*f*t)]^n$,

$R_{2y}(t) = r_0^n * e^{[-n*2^{p+2}*\sin^2(\pi*f*t/2^p)]} * [sin(2*\pi*k*f*t)]^n$,

Parametric curve:

$e = Curve(R_{2x}(t), R_{2y}(t), t, -2*\pi, 2*\pi)$,

$R_2(f) = r_0^n * e^{[-n*2^{p+2}*\sin^2(\pi*f*t/2^p)]}$,

$R_{2x}(f) = r_0^n * e^{[-n*2^{p+2}*\sin^2(\pi*f*t/2^p)]} * [cos(2*\pi*k*f*t)]^n$,

$R_{2y}(f) = r_0^n * e^{[-n*2^{p+2}*\sin^2(\pi*f*t/2^p)]} * [sin(2*\pi*k*f*t)]^n$,

Parametric curve:

$e = Curve(R_{2x}(f), R_{2y}(f), f, -2*\pi, 2*\pi)$.

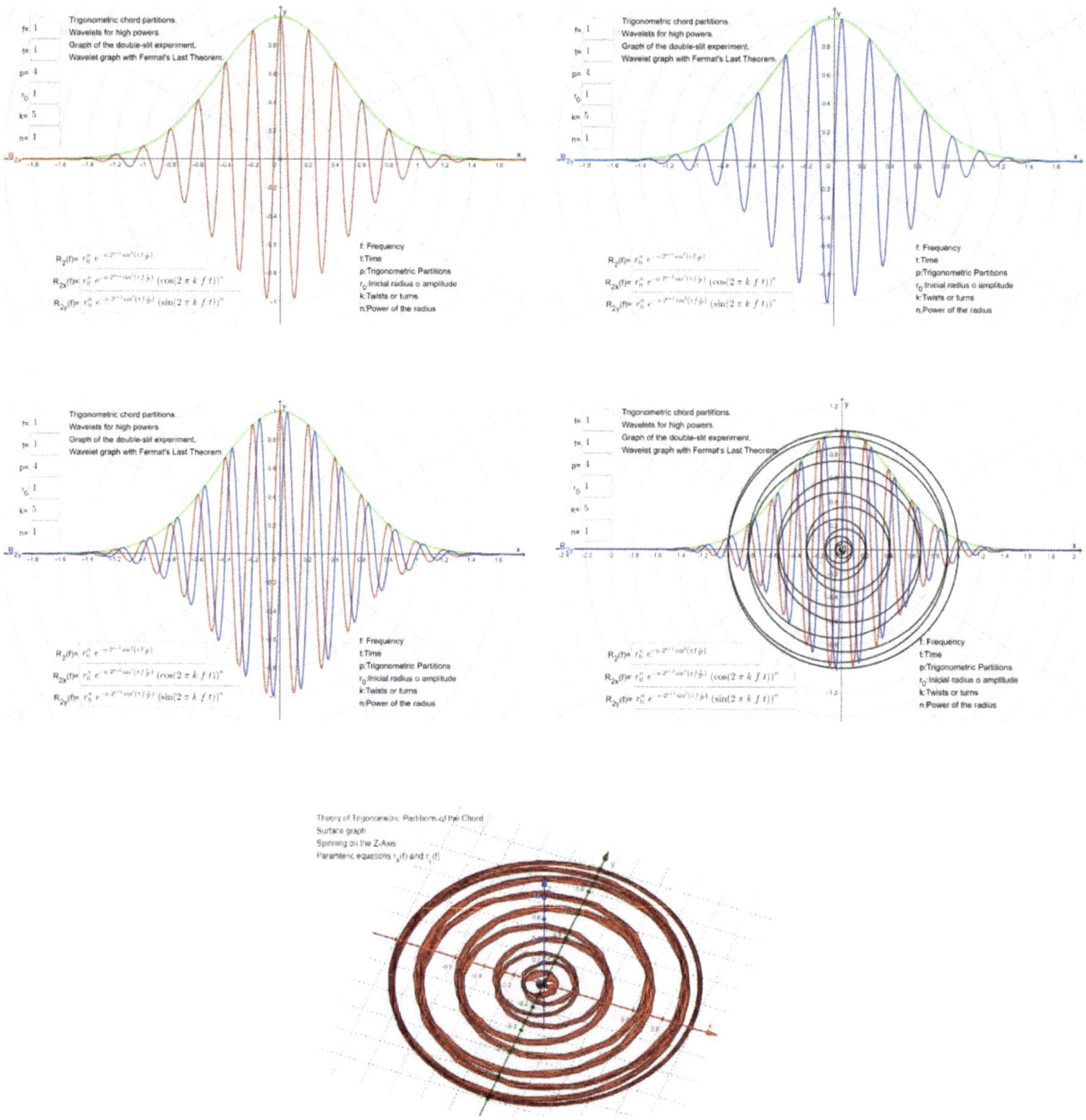

Figure 1.5: Graphs of the uncertainty principle with trigonometric partitions.

Wavelets of nth powers:

These wavelets of nth powers are based on the Pythagorean equation $r^2 = x^2 + y^2$, although I don't sum the components x^2 and y^2. Therefore, each equation of the trigonometric chord partitions is raised to the nth power, similar to Fermat's Last Theorem. Similarly, a wide variety of wavelets can be generated, such as the graph generated in the double-slit experiment. In these equations, the domains are bounded to small values to prevent GeoGebra from slowing down due to the large number of calculations used by the program. When these values are increased to large values, interesting graphs can be observed. When we change each variable, we can analyze the effects that each variable causes in the equations of the trigonometric chord partitions. This simple methodology helps us analyze physical cases like the double- slit experiment. Additionally, multiple graphs of the double- slit experiment and wavelets can be generated. This new methodology of trigonometric partitions allows us to study wavelets at both low and high frequencies. In the book I have written, you can learn about the various applications where the theory of trigonometric partitions can be used. This tool is similar to the Fourier series theory, but it is not the Fourier series. With this new partition tool, the number of known wavelets has increased. Initially, when I started developing this theory, I applied the concepts of spiral angles to the double- slit experiment, where I had successful results. Later, I realized that I had radii and could also apply trigonometric partitions. In a social media video, I recorded, observe what happens when prime and odd numbers are used and how the wavelet shape changes. Additionally, with the Theory of Trigonometric Partitions, we can study the harmonics that deform the sine and cosine waves and generate their graphs. With the first and second derivatives of the trigonometric partitions of the chord can also generate duplex or double wavelets. Graphs of the equations of the trigonometric partitions of the chord as a function of frequency and time:

$$R_2(t) = r_0^n * e^{[-n*2^{p+2}*\sin^2(\pi*f*t/2^p)]},$$

$R_{2x}(t) = r_0^n * e^{[-n*2^{p+2}*\sin^2(\pi*f*t/2^p)]} * [cos(2*\pi*k*f*t)]^n,$

$R_{2y}(t) = r_0^n * e^{[-n*2^{p+2}*\sin^2(\pi*f*t/2^p)]} * [sin(2*\pi*k*f*t)]^n,$

Parametric curve:

$e = Curve(R_{2x}(t), R_{2y}(t), t, -2*\pi, 2*\pi),$

$R_2(f) = r_0^n * e^{[-n*2^{p+2}*\sin^2(\pi*f*t/2^p)]},$

$R_{2x}(f) = r_0^n * e^{[-n*2^{p+2}*\sin^2(\pi*f*t/2^p)]} * [cos(2*\pi*k*f*t)]^n,$

$R_{2y}(f) = r_0^n * e^{[-n*2^{p+2}*\sin^2(\pi*f*t/2^p)]} * [sin(2*\pi*k*f*t)]^n,$

Parametric curve:

$e = Curve(R_{2x}(f), R_{2y}(f), f, -2*\pi, 2*\pi),$

A:This graph is for the light: X Coordinate.

B:This graph is for the darkness: Y Coordinate.

C and D: Charts combination: X and Y Coordinate.

E:Surface graph: The red line is the parametric curve

.

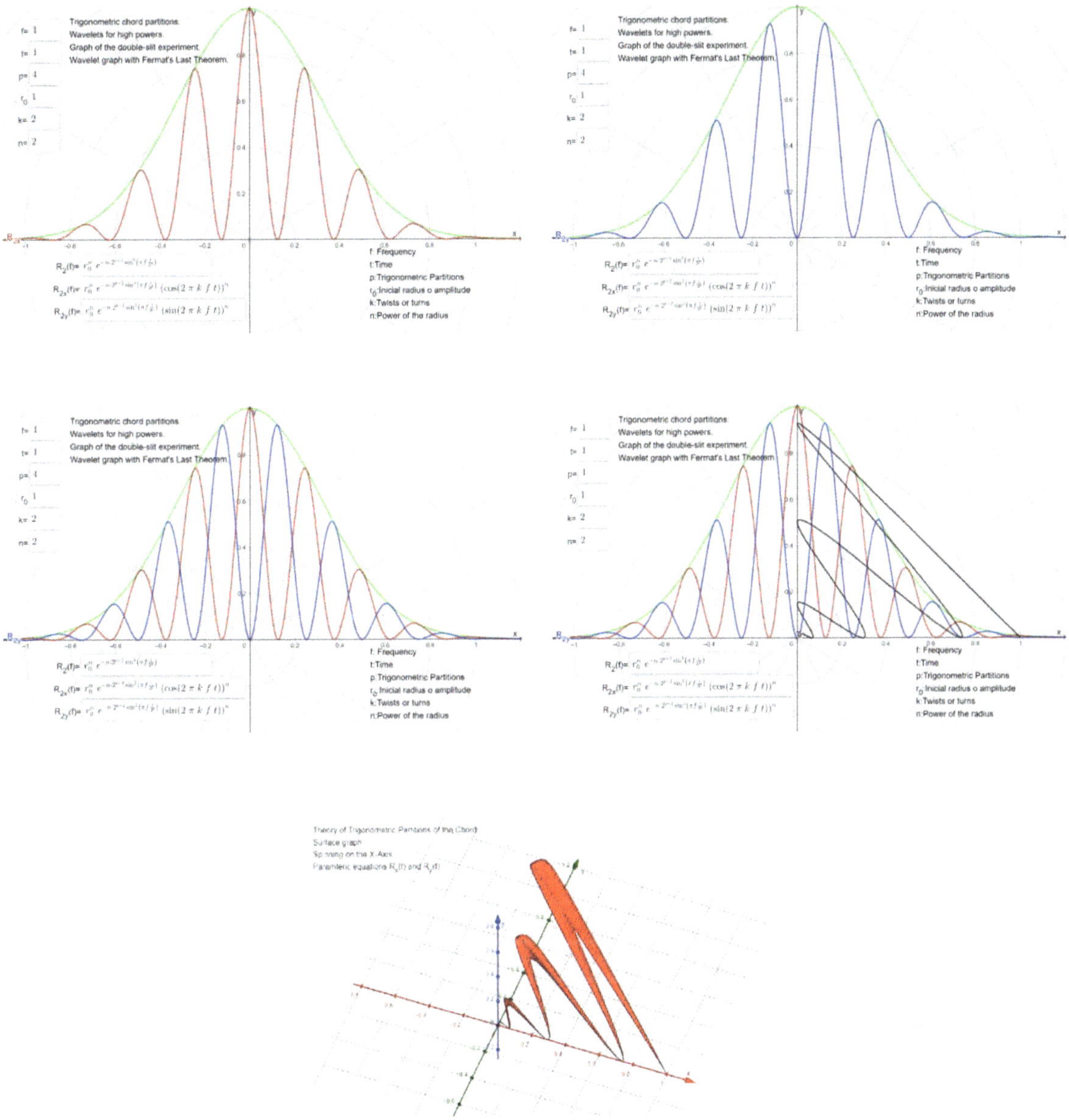

Figure 1.6: Graphs of the double slit experiment with trigonometric partitions.

I have written a book about the Trigonometric Partitions. You can buy and learn all about the application of this new method. Visit my social networks where I talk more about this topic of the Theory of Spiral Angles, Spirals, and Trigonometric Partitions. In these videos where I talk about some topics that are not written and explained in the book.

Theory of Spiral Angles, Spirals, and Trigonometric Partitions: It is the theory that can study and analyze periodic and non-periodic functions, without limits of periods, frequency, angular speed, and time, therefore, it was developed in a circular way; similar to the waves that are generated in water by potential energy. There are seven mathematical equations that help us to study in an integral way, the components of a circle, which is continuously divided into halves to infinity. The part that continuously divides the circle to study it by sections is called; Trigonometric Partitions which can connect all the planes into one; as are the planes of real and imaginary numbers. Therefore, for more knowledge of these theories you should read the book that I have written in Barker and Jules. Of which, I own the copyright. Library Of Congress Copyrights Control Number: 1 – 12354195611 at USA. With the Theory of Trigonometric Partitions and Spiral Angles, I have created a new technique to study Prime Numbers and Riemann's Z Function or Hypothesis´s Riemann. And solve this problem from a new mathematics point of view. The technique developed is the following; calculate the trigonometric partitions by trial and error approximations up to the approximation of 6 digits or those that are desired. Prime Numbers are nested or concatenated with each other, as are non-prime numbers; even and odd, etc. For example, the following prime numbers: 3; I will call it the Previous Prime Number and 5; I will call it the Posterior Prime Number, and its subtraction of 5-3=2; I will call it the length module of the Prime Numbers or $\Delta L = \Delta r = \Delta P$. It is the longest module of the Prime Numbers that connects to the equation of the Riemann's Hypothesis. Both equations that I have developed; are one of the common points of both formulas and it is for this reason that both equations are linked.

The Prime Numbers equation uses this trigonometric partition, the spiral angle, the Previous Prime Number, and the length module of the Prime Numbers to calculate the Posterior Prime Number. Initially, this equation of Prime Numbers can be written in its general form; in which you must be very careful with the formula that determines the Spiral Angles to be used in the mathematical formula. Later, with this same formula, it can be analyzed with the constant spiral angle as I have taught in the Theory of Spiral Angles, Spirals, and Trigonometric Partitions. The calculation of the Trigonometric Partitions depends on the initial Spiral Angle to be studied or on the selected formula to determine the Spiral Angles. For example, you can start with an angle of 45°,60° or 90° degrees constantly, your results of the trigonometric partitions will be different for both degrees initially chosen. When this technique is chosen, the Riemann Z Function equation is capable of generating graphs equal to those of the Riemann Z Function. One of the questions that we always ask ourselves about all Prime Numbers is: ¿How do we discriminate all the Prime Numbers from the real numbers? The tool that helps us study prime numbers and their behavior is their length module and the height **f** on the Y axis. The following graph is a table with 27 points where the subtraction of the subsequent posterior prime numbers minus the previous prime number is the length module of the prime numbers $\Delta L = \Delta r = \Delta P$. The table is generated from the prime numbers 2 to 101. Which can be extended to all Prime Numbers. This same pattern of this graph is the one that will be pursued with the equation of the prime numbers later, where we will observe how the pattern is maintained with the equation where the trigonometric partitions are searched for by trial and error and with this result the prime can be obtained posterior in the real plane, with the values initially selected; which are very important to carry out a correct analysis of Prime Numbers.

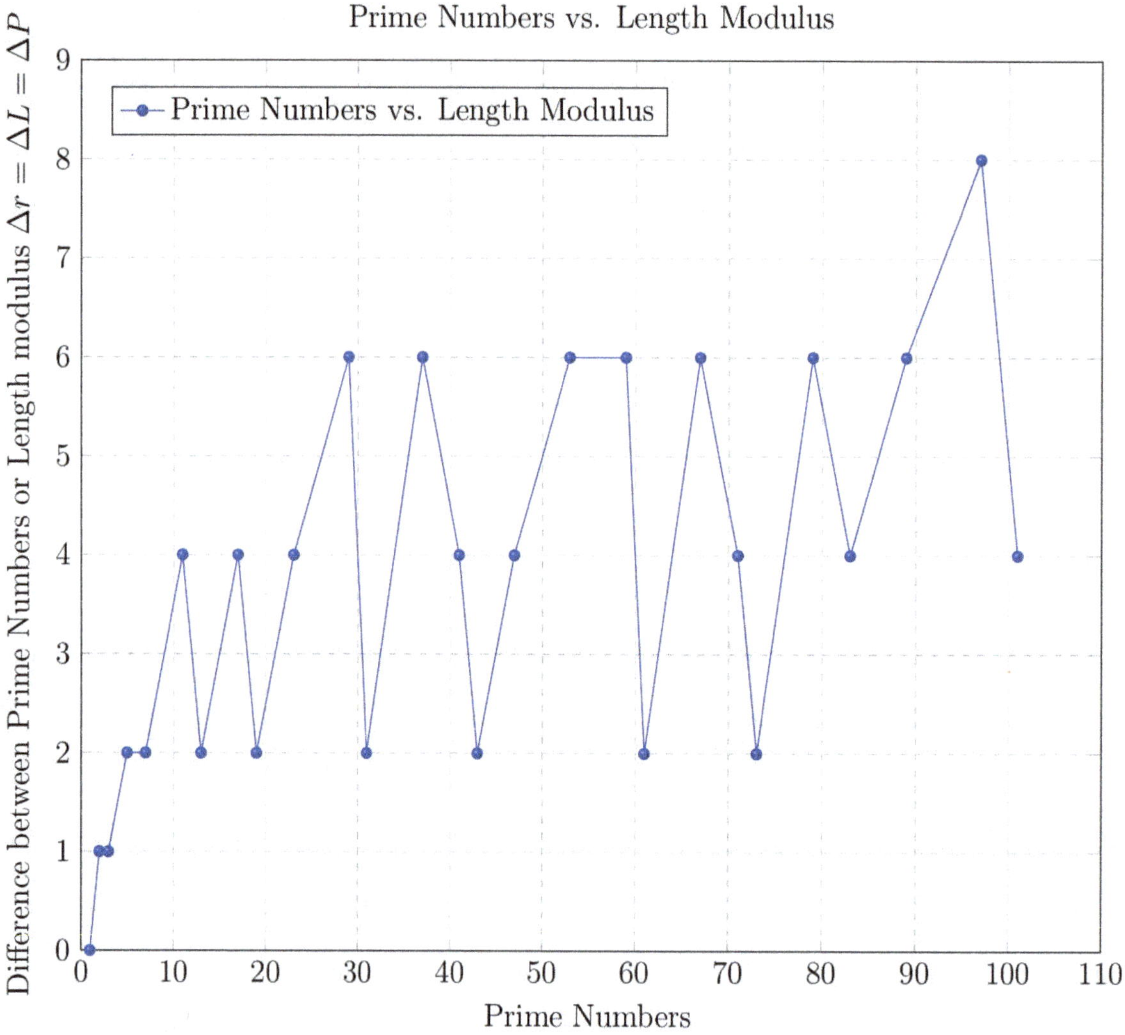

Figure 1.7: Graph showing the relationship between prime numbers and their length modulus, represented as the difference between consecutive primes ($\Delta r = \Delta L = \Delta P$).

With the Theory of Spiral Angles, Spirals, and Trigonometric Partitions, Analytical Geometry can be studied in a more effective way. Since the functions of the Prime Numbers and the Riemann Z Function have different behavior in the real plane and the imaginary plane. In the graph above we can study the slopes of the change from a prime number before to the one after it. Trigonometric partitions are positive when their angular velocity is counterclockwise and they are negative when their angular velocity is clockwise. The analysis starts with the unit circle of 45°,60° and 90° degrees or any other starting angle that you want to use. The original equation of the Riemann Z Function in the 2D plane is capable of generating graphs.

equal to the Riemann Z Function with f=0, when we obtain the two equations for the X and Y polar coordinates, its equation is invisible in the GeoGebra program. You must mark the trace to be able to see the behavior of the two polar coordinates in the 2D plane;which are functions of points . These same polar coordinates of X and Y, in the 3D plane, when we do the integration of the solid by revolution; generate arcs or circles depending on the spiral angle and the value of k turns; initially chosen. An important point: when you want to calculate large prime numbers with greater precision, you must choose more decimal places for rounding, to calculate their partitions and their non-trivial zeros, since it depends on knowing more precisely what their non-trivial zeros are. In the Cartesian plane, theyare the Trigonometric Partitions and their Spiral Angles, which help us to place the Prime Numbers in the correct position, to analyze their behavior, either linearly when the initial spiral angle is constant or circular when the spiral angle uses an equation with a variable length Therefore, in both cases , we will have different graphs that show the behavior of each equation of the Riemann Z Function or Prime Numbers. The followingdiagram shows us how we can place the Prime Numbers in a Cartesian plane, to apply the Theory of Trigonometric Partitions and their Spiral Angles. And demonstrate what is the equation of the Prime Numbers in the real and Imaginary plane. $\Delta L = \Delta r = r - r$, The interest is the radii that are found along the X axis, which will represent the Prime Numbers located in the X axis. r_2 and r_4 are the radii located radially, each one with its flat circular waves, which will connect us with the equations of the Trigonometric Partitions. Defining by the Pythagorean theorem; $r_2 = \sqrt{(r_1^2 + f^2)}$. Therefore, the radius has two roots. Defining by the Theory of Trigonometric Partitions of the growth of the radius; With the radius in compression; $r_2 = r_1 * e^{2(p-1)*\sin(\theta/2^{p-1})-\theta}$; This equation is in radians. Defining by the Pythagorean theorem; $r_4 = \sqrt{(r_3^2 + f^2)}$;Therefore, the radius has two roots. Defining by the Theory of Trigonometric Partitions of the growth of the radius; With the radius in compression; $r_4 = r_3 * e^{2(p-1)*\sin(\theta/2^{p-1})-\theta}$;This equation is in radians. Substitute in the equation of the length modules of the Prime Numbers; $\Delta L = \Delta r = r_3 - r_1$ And solving the equations of the Trigonometric Partitions to r_3, r_1, are obtained;

$r_1 = r_2 * e^{\theta - 2^{(p-1)} * \sin(\theta/2^{(p-1)})}, r_3 = r_4 * e^{\theta - 2^{(p-1)} * \sin(\theta/2^{(p-1)})}$.

$\Delta L = r_3 - r_1 = r_4 * e^{\theta - 2^{(p-1)} * \sin(\theta/2^{(p-1)})} - r_2 * e^{\theta - 2^{(p-1)} * \sin(\theta/2^{(p-1)})} = (r_4 - r_2) * e^{\theta - 2^{(p-1)} * \sin(\theta/2^{(p-1)})}$.

Substituting the previous equation to; $r_4 = \sqrt{(r_3^2 + f^2)}$ and $r_2 = \sqrt{(r_1^2 + f^2)}$

Therefore;$\Delta L = (r_4 - r_2) * e^{\theta - 2^{(p-1)} * \sin(\theta/2^{(p-1)})} = [\sqrt{(r_3^2 + f^2)} - \sqrt{(r_1^2 + f^2)}] * e^{\theta - 2^{(p-1)} * \sin(\theta/2^{(p-1)})}$.

What we require is the Posterior Prime Number r_3 as a function of the Previous Prime number r_1 and the partitions **p**, with their spiral angles θ, with a height **f** of the Y axis, which can be variable or constant; depending on the initial angle that we choose to analyze the real or complex plane. The height variable **f** on the Y axis; helps us to analyze in matrix form the equation of Prime Numbers and the Riemann Hypothesis. Solving for the posterior radius the equation of the Prime Numbers in general form would be: $\sqrt{(r_3^2 + f^2)} = \Delta L * e^{2^{(p-1)} * \sin(\theta/2^{p-1}) - \theta} + \sqrt{(r_1^2 + f^2)}$, Squaring both sides of the equation gives; $r_3^2 + f^2 = [\Delta L * e^{2^{(p-1)} * \sin(\theta/2^{p-1}) - \theta} + \sqrt{(r_1^2 + f^2)}]^2$,

$r_3^2 = [\Delta L * e^{2^{(p-1)} * \sin(\theta/2^{p-1}) - \theta} + \sqrt{(r_1^2 + f^2)}]^2 - f^2$.

Finally, the Prime Posterior number will be equal to the posterior radius:

$r_3(p) = \sqrt{[\Delta L * e^{2^{(p-1)} * \sin(\theta/2^{p-1}) - \theta} + \sqrt{(r_1^2 + f^2)}]^2 - f^2}$,

remembering that $\Delta L = \Delta r = r_3 - r_1$.

$r_3(p) = \sqrt{[\Delta r * e^{2^{(p-1)} * \sin(\theta/2^{p-1}) - \theta} + \sqrt{(r_1^2 + f^2)}]^2 - f^2}$; which has two solutions, one positive and one negative since it is a square root.

The following graph is the schematic where the obtained equations can be analyzed.

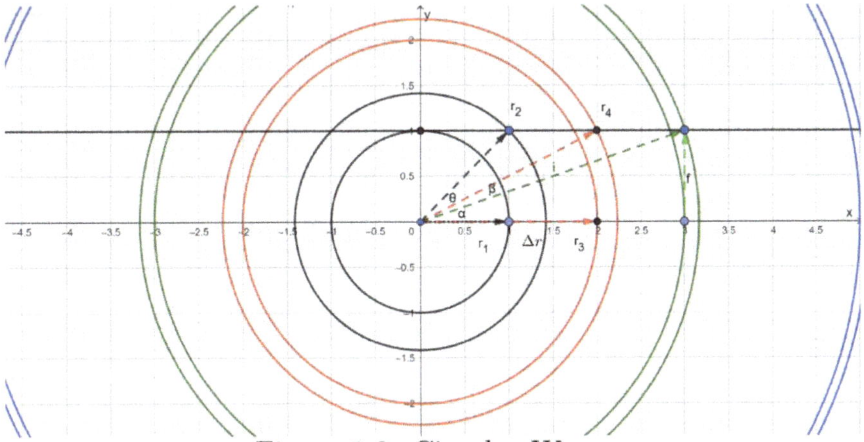

Figure 1.8: Circular Waves

If we analyze this same equation with a constant theta initial spiral angle; the initial values are the most important values to obtain the correct Trigonometric Partitions by trial and error or by estimation, with the height **f**; which is the initial value to determine the initial spiral angle. The value f can be any value. Examples f = 0,1,$\sqrt{2}/2$, $\sqrt{3}/2$, ... etc. The Trigonometric Partitions will be different for each of these values of **f**. However, their behavior on the charts will be the same. With this equation we will follow the behavior of the Prime Numbers in relation to the length module of the Prime Numbers: ΔL. And a great diversity of graphics that can be obtained in Excel spreadsheet on the real plane. I will also use the GeoGebra program to observe the behavior of the equations and perform calculations in the real and complex plane with different analyses, such as; constant and variable spiral angles. Likewise, **f** variable and constant, to observe the effects that each variable causes in the equations of the Prime Numbers, and the Riemann Hypothesis.

$\theta = 45^\circ = \pi/4 = \pi/2^2; r_3(p) = \sqrt{[\Delta L * e^{2(p-1)*\sin(\pi/2(p+1))-\pi/4} + \sqrt{(r_1^2 + f^2)}]^2 - f^2}.$

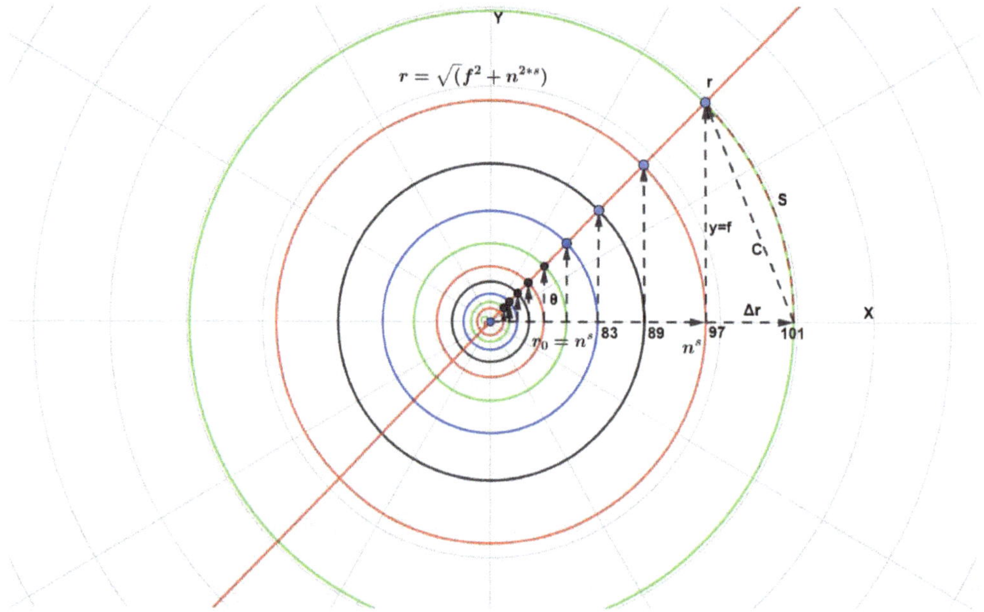

Qty	Prime Numbers	Δr=ΔL
1	1	0
2	2	1
3	3	1
4	5	2
5	7	2
6	11	4
7	13	2
8	17	4
9	19	2
10	23	4
11	29	6
12	31	2
13	37	6
14	41	4
15	43	2
16	47	4
17	53	6
18	59	6
19	61	2
20	67	6
21	71	4
22	73	2
23	79	6
24	83	4
25	89	6
26	97	8
27	101	4

Figure 1.9: Prime Numbers vs. Length Modulus. The first image shows a graphical representation, while the second provides a tabular comparison of prime numbers against their respective length modulus.

1.2 Definitions Of The Equations Of Prime Numbers In The Real Plane.

The General Form of the Prime Numbers equation for $\Delta L, \theta, p, r, f, s$ variables. The value of the partition **p** can be calculated first by trial and error, with the constant spiral angle, this same partition can be used with the variable spiral angle, since the **p** partitions are lines that pass through the center, cutting the Cartesian plane and they can be the same partitions with different angles, since these partitions can work with plane circular waves in the Cartesian plane.

Using this same methodology and the different equations of trigonometric partitions, we can create a wide variety of tables, calculated through trial and error, for each trigonometric partition **p**, which is related to each prime number. Subsequently, we can use these data of **p** in various mathematical equations to study the behavior of the graph of the Riemann

Zeta function.

$r_3(p) = \sqrt{[\Delta L * e^{2^{(p-1)}*\sin(\theta/2^{(p-1)})-\theta} + \sqrt{(r_1^2+f^2)}]^2 - f^2}$, remembering that $\Delta L = \Delta r = r_3 - r_1$, $r_3(p) = \sqrt{[\Delta r * e^{2^{(p-1)}*\sin(\theta/2^{(p-1)})-\theta} + \sqrt{(r_1^2+f^2)}]^2 - f^2}$;which has two solutions, one positive and one negative since it is a square root.Transformed the general form of the Prime Numbers equation, for θ and s=1 constants and ΔL, p, r_1, f variables. The initial angle can be any value that is desired to start the analysis of the behavior of the Prime Numbers equation. This same concept will be applied later for the Riemann Hypothesis equation, which is linked by the value of the length module of the Prime Numbers and the height **f** of the Y axis. This same technique can be used for even numbers, and odd, or any other analysis that you want to develop with the numbers. The first difference in each of these two equations at the time of generating the graph is their beginnings and the values of the subsequent prime number: r_3 Consequently, the behavior of their graphs is the same when studying the partitions vs. the modules of length. **K** is the number of turns or twists that can be generated with the equation.

With the following equations; The following tables will be generated in Excel spreadsheet to study the behavior of the Prime Numbers versus and their radial growths $\Delta r = \Delta L = r_3 - r_1$. And calculate the Trigonometric Partition that connects with each Prime Number in the Cartesian plane.

$\theta = 45° = \pi/4 = \pi/2^2$;

$r_3(p) = \sqrt{[\Delta L * e^{[2^{(p-1)}*\sin(\pi*k/2^{(p+1)})-\pi*k/4]} + \sqrt{(r_1^2+f^2)}]^2 - f^2]}$

$\theta = 60° = \pi/3$

; $r_3(p) = \sqrt{[\Delta L * e^{[2^{(p-1)}*\sin(\pi*k/(3*2^{(p+1)}))-\pi*k/3]} + \sqrt{(r_1^2+f^2)}]^2 - f^2}$.

With a constant starting angle; $\theta = 45° = \pi/4$; The Trigonometric Partitions are calculated by trial and error; $r_3(p) = \sqrt{[\Delta L * e^{[2^{(p-1)}*\sin(\pi*k/2^{(p+1)})-\pi*k/4]} + \sqrt{(r_1^2+f^2)}]^2 - f^2}$.

The following tables that have been generated in Excel spreadsheet are where the calculations of the trigonometric partitions **p** are shown by trial and error. The variables m and **q** have been placed just to make a large variety of the exponent **s**. In these tables in Excel spreadsheet; $s = \frac{m}{q}$. The exponent **s** represents a mathematical equation P^s, the power **s** of the prime number **P** in the real plane has the property of expanding or shrinking the graph; in the complex plane it produces rotation, if **s** is changed, so do trigonometric partitions. In the calculated posterior prime column is where the prime number with these same values is calculated. 6 decimal places are used in the precision of the rounding about the trigonometric partition, this 6-digit precision can be increased for large prime numbers, and thus obtain non-trivial zeros with rounding of more than 6 decimal places and more precise. This is just the equation of the first quadrant; $r_3(p) = \sqrt{[\Delta L * e^{[2^{(p-1)}*\sin(\pi*k/2^{(p+1)})-\pi*k/4]} + \sqrt{(r_1^2+f^2)}]^2 - f^2}$.

, but you can also obtain the one from quadrant 4 which is the negative equation, since it is a square root it has two solutions.$r_3(p) = -\sqrt{[\Delta L * e^{[2^{(p-1)}*\sin(\pi*k/2^{(p+1)})-\pi*k/4]} + \sqrt{(r_1^2+f^2)}]^2 - f^2}$.

The following tables can be used with the general equation of the Riemann Zeta function and observe all the graphs that are generated. Likewise, tables can be generated with f=0, which is equivalent to generating an equation on the X-axis and creating simpler equations that generate the graph of the Riemann Zeta function, which can be created with real and complex numbers.

If we create tables with different initial angles with f=0 and use them in the general equations of the Riemann Z function of trigonometric partitions; We are always going to get the graph of the Riemann Z function of different sizes.

Table 1.3: Prime Numbers Formula for +45 Degrees.

Factc	f=	1	m=	0	Q=	1	k=	1	s=	0
Qty	**Posterior prime number partitions** p_1	**Previo Prime Number** r_1	**Modul** ΔL	**Modul** $\sum \Delta L$	**Posterior prime number** r_3	**Spiral Angle** θ **Rads**	**Spiral Angle** θ **Degs**	**Calculated Posterior Prime Number** r_3	**Coord X Real Posterior Prime Number** r_{x3}	**Coord Y Real Posterior Prime Number** r_{y3}
1	*1*	*1*	*0.000*	0	1	0.785398	45	1.000000	0.707107	0.707107
2	*0.061199*	*1.00*	*0.000*	*0*	1	0.785398	45	1.000000	0.707107	0.707107
3	*0.40928*	*1*	*0.000*	*0*	1	0.785398	45	1.000000	0.707107	0.707107
4	*0.71501*	*1*	*0.000*	*0*	1	0.785398	45	1.000000	0.707107	0.707107
5	*1.00265*	*1*	*0.000*	*0*	1	0.785398	45	1.000000	0.707107	0.707107
6	*1.27646*	*1*	*0.000*	*0*	1	0.785398	45	1.000000	0.707107	0.707107
7	*1.4927*	*1*	*0.000*	*0*	1	0.785398	45	1.000000	0.707107	0.707107
8	*1.64649*	*1*	*0.000*	*0*	1	0.785398	45	1.000000	0.707107	0.707107
9	*1.7807*	*1*	*0.000*	*0*	1	0.785398	45	1.000000	0.707107	0.707107
10	*1.8882*	*1*	*0.000*	*0*	1	0.785398	45	1.000000	0.707107	0.707107
11	*2.03875*	*1*	*0.000*	*0*	1	0.785398	45	1.000000	0.707107	0.707107
12	*2.145*	*1*	*0.000*	*0*	1	0.785398	45	1.000000	0.707107	0.707107
13	*2.2324*	*1*	*0.000*	*0*	1	0.785398	45	1.000000	0.707107	0.707107
14	*2.3322*	*1*	*0.000*	*0*	1	0.785398	45	1.000000	0.707107	0.707107
15	*2.386*	*1*	*0.000*	*0*	1	0.785398	45	1.000000	0.707107	0.707107
16	*2.435*	*1*	*0.000*	*0*	1	0.785398	45	1.000000	0.707107	0.707107
17	2.5098	1	0.000	0	1	0.785398	45	1.000000	0.707107	0.707107
18	2.59135	1	0.000	0	1	0.785398	45	1.000000	0.707107	0.707107
19	2.642	1	0.000	0	1	0.785398	45	1.000000	0.707107	0.707107
20	2.6876	1	0.000	0	1	0.785398	45	1.000000	0.707107	0.707107
21	2.742	1	0.000	0	1	0.785398	45	1.000000	0.707107	0.707107
22	2.773	1	0.000	0	1	0.785398	45	1.000000	0.707107	0.707107
23	2.811	1	0.000	0	1	0.785398	45	1.000000	0.707107	0.707107
24	2.8576	1	0.000	0	1	0.785398	45	1.000000	0.707107	0.707107
25	2.9	1	0.000	0	1	0.785398	45	1.000000	0.707107	0.707107
26	2.956	1	0.000	0	1	0.785398	45	1.000000	0.707107	0.707107
27	3.002	1	0.000	0	1	0.785398	45	1.000000	0.707107	0.707107

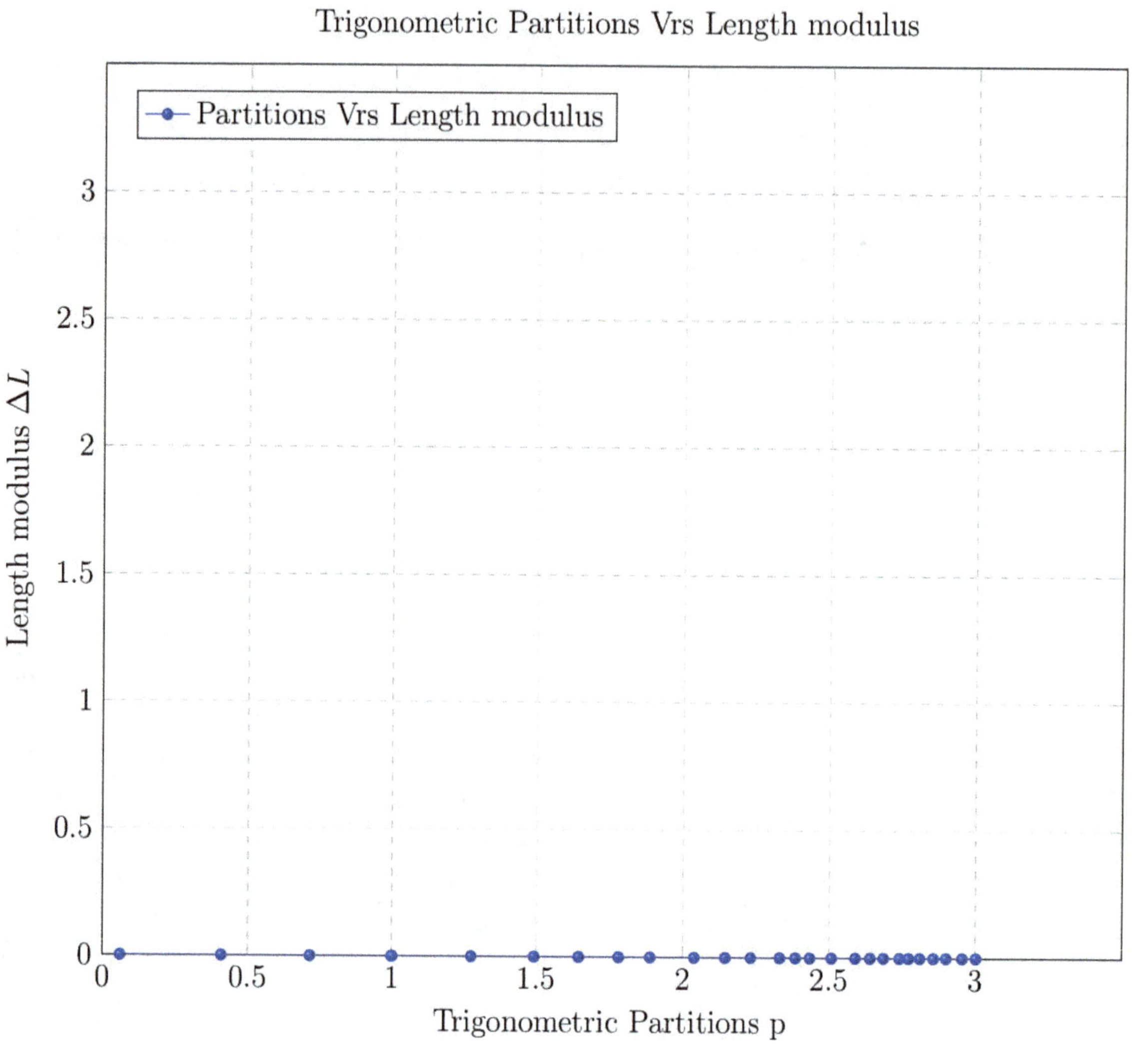

Figure 1.10: Trigonometric Partitions Vrs Length modulus,at 45 degrees, s=0,f=1.

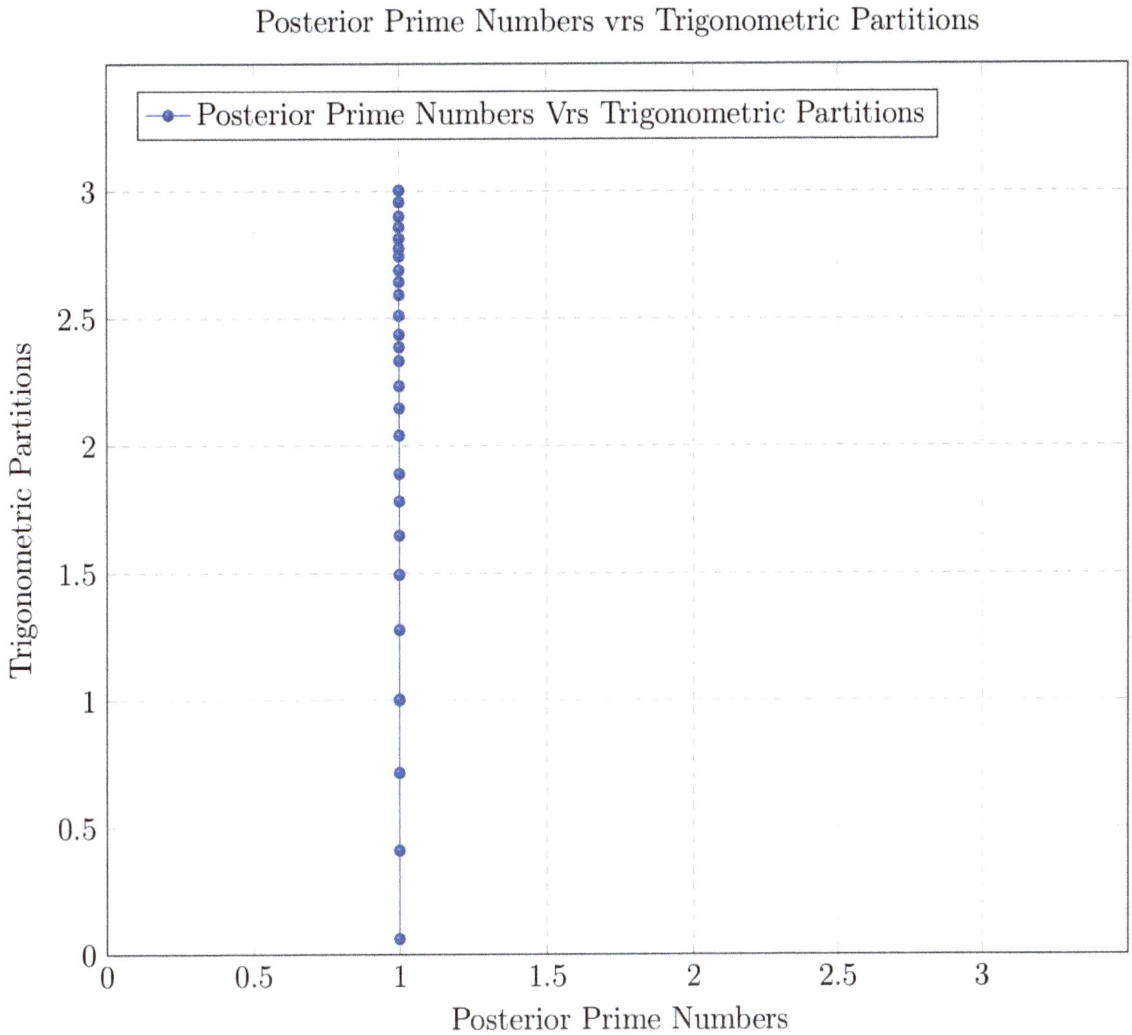

Figure 1.11: Posterior Prime Numbers Vrs Trigonometric Partitions, at 45 degrees, s=0,f=1.

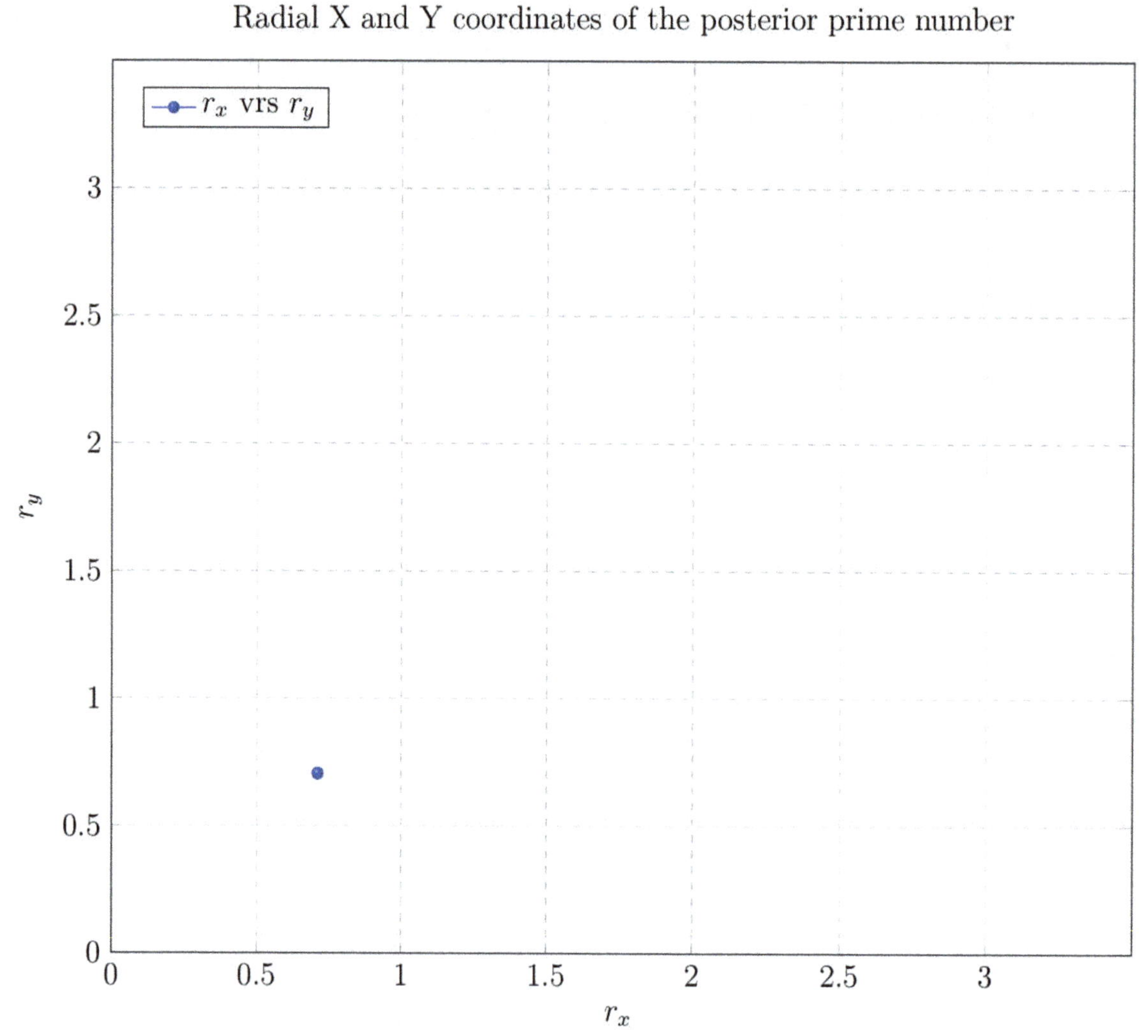

Figure 1.12: Radial X and Y coordinates of the posterior prime number, at 45 degrees, s=0,f=1.

Table 1.4: Prime Numbers Formula for +45 Degrees.

Facto	f=	1	m=	1	Q=	2	k=	1	s=	0.5
Qty	**Posterior prime number partitions** p_1	**Previou Prime Number** r_1	**Modul** ΔL	**Modul** $\sum \Delta L$	**Posterior prime number** r_3	**Spiral Angle** θ **Rads**	**Spiral Angle** θ **Degs**	**Calculated Posterior Prime Number** r_3	**Coord X Real Posterior Prime Number** r_{x3}	**Coord Y Real Posterior Prime Number** r_{y3}
1	1	1	0.000	0	1	0.785398	45	1.000000	0.707107	0.707107
2	0.061199	1.00	0.414	0.4142	1.414214	0.785398	45	1.414214	1.000000	1.000000
3	0.40928	1.41421	0.318	0.7321	1.732051	0.785398	45	1.732051	1.224745	1.224745
4	0.71501	1.73205	0.504	1.2361	2.236068	0.785398	45	2.236068	1.581139	1.581139
5	1.00265	2.23607	0.410	1.6458	2.645751	0.785398	45	2.645751	1.870829	1.870829
6	1.27646	2.64575	0.671	2.3166	3.316625	0.785398	45	3.316625	2.345208	2.345208
7	1.4927	3.31662	0.289	2.6056	3.605551	0.785398	45	3.605551	2.549509	2.549509
8	1.64649	3.60555	0.518	3.1231	4.123106	0.785398	45	4.123106	2.915476	2.915476
9	1.7807	4.12311	0.236	3.3589	4.358899	0.785398	45	4.358899	3.082207	3.082207
10	1.8882	4.3589	0.437	3.7958	4.795832	0.785398	45	4.795832	3.391166	3.391166
11	2.03875	4.79583	0.589	4.3852	5.385165	0.785398	45	5.385165	3.807887	3.807887
12	2.145	5.38516	0.183	4.5678	5.567764	0.785398	45	5.567764	3.937004	3.937004
13	2.2324	5.56776	0.515	5.0828	6.082763	0.785398	45	6.082765	4.301164	4.301164
14	2.3322	6.08276	0.320	5.4031	6.403124	0.785398	45	6.403124	4.527693	4.527693
15	2.386	6.40312	0.154	5.5574	6.557439	0.785398	45	6.557439	4.636809	4.636809
16	2.435	6.55744	0.298	5.8557	6.855655	0.785398	45	6.855655	4.847680	4.847680
17	2.5098	6.85565	0.424	6.2801	7.280110	0.785398	45	7.280110	5.147815	5.147815
18	2.59135	7.28011	0.401	6.6811	7.681146	0.785398	45	7.681146	5.431390	5.431390
19	2.642	7.68115	0.129	6.8102	7.810250	0.785398	45	7.810250	5.522681	5.522681
20	2.6876	7.81025	0.375	7.1854	8.185353	0.785398	45	8.185353	5.787919	5.787919
21	2.742	8.18535	0.241	7.4261	8.426150	0.785398	45	8.426150	5.958188	5.958188
22	2.773	8.42615	0.118	7.544	8.544004	0.785398	45	8.544004	6.041523	6.041523
23	2.811	8.544	0.344	7.8882	8.888194	0.785398	45	8.888194	6.284902	6.284902
24	2.8576	8.88819	0.222	8.1104	9.110434	0.785398	45	9.110434	6.442050	6.442050
25	2.9	9.11043	0.324	8.434	9.433981	0.785398	45	9.433981	6.670832	6.670832
26	2.956	9.43398	0.415	8.8489	9.848858	0.785398	45	9.848858	6.964194	6.964194
27	3.002	9.84886	0.201	9.0499	10.049876	0.785398	45	10.049876	7.106336	7.106336

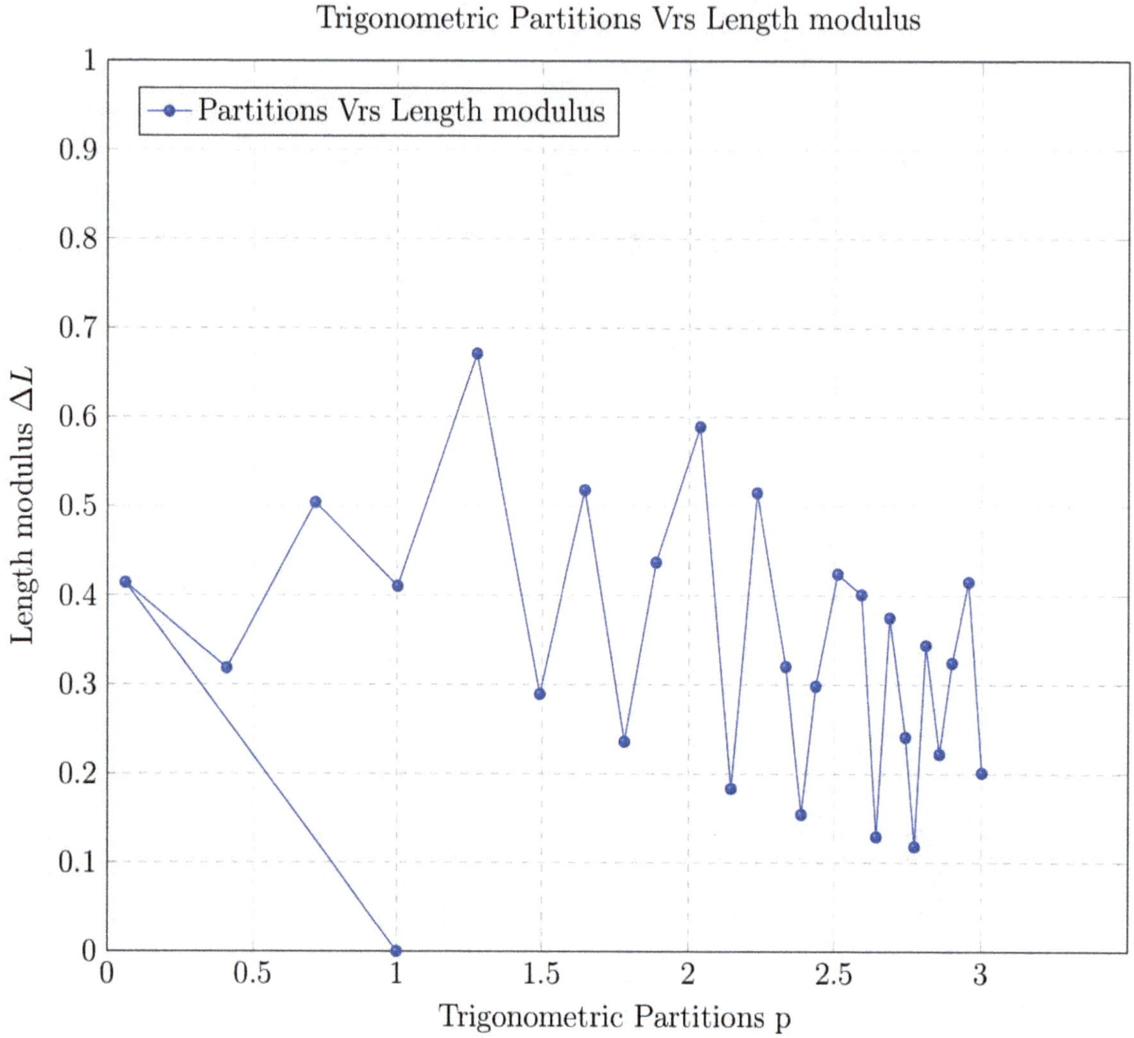

Figure 1.13: Trigonometric Partitions Vrs Length modulus, at 45 degrees, s=0.5, f=1.

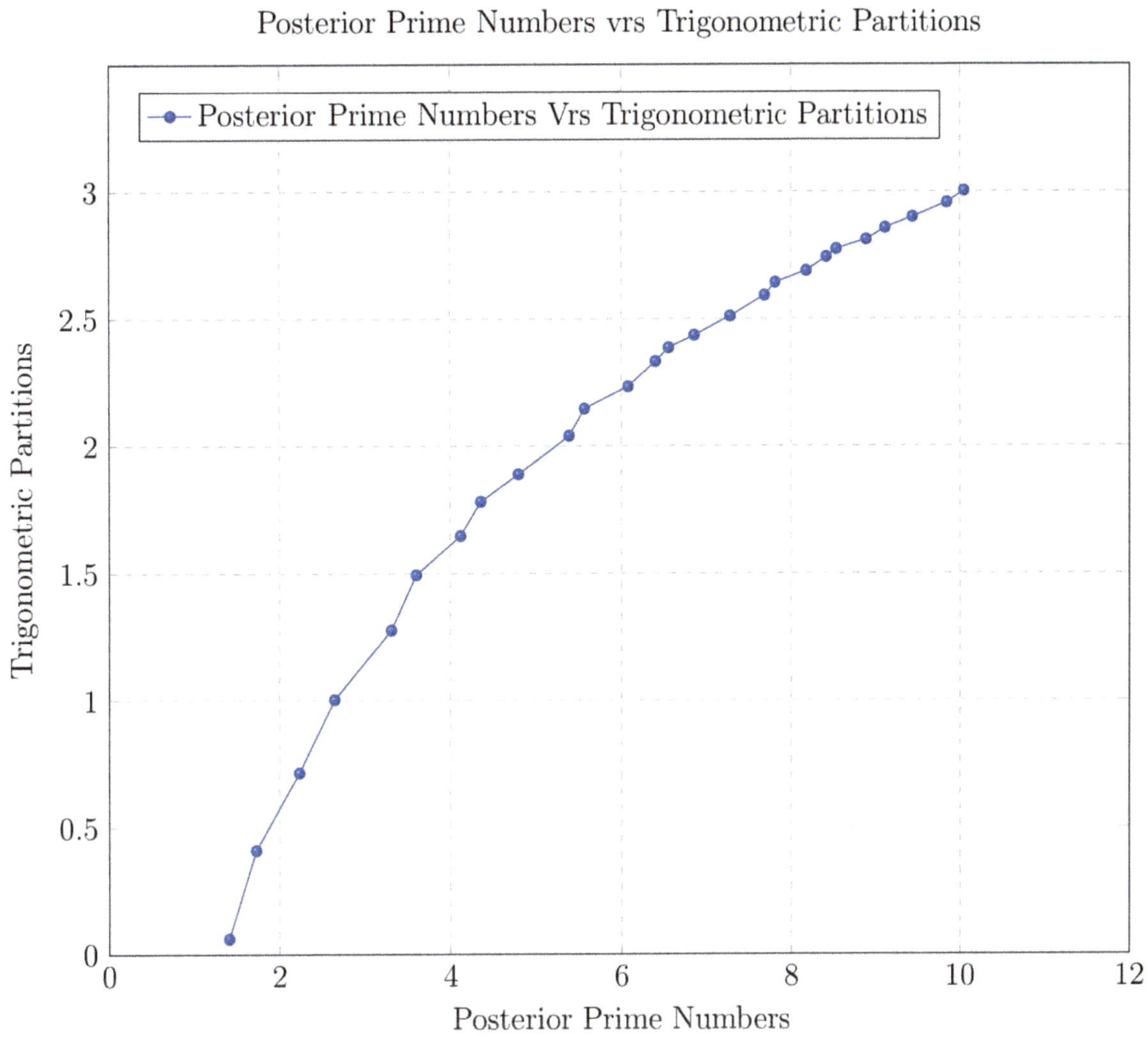

Figure 1.14: Posterior Prime Numbers Vrs Trigonometric Partitions, at 45 degrees, s=0.5, f=1.

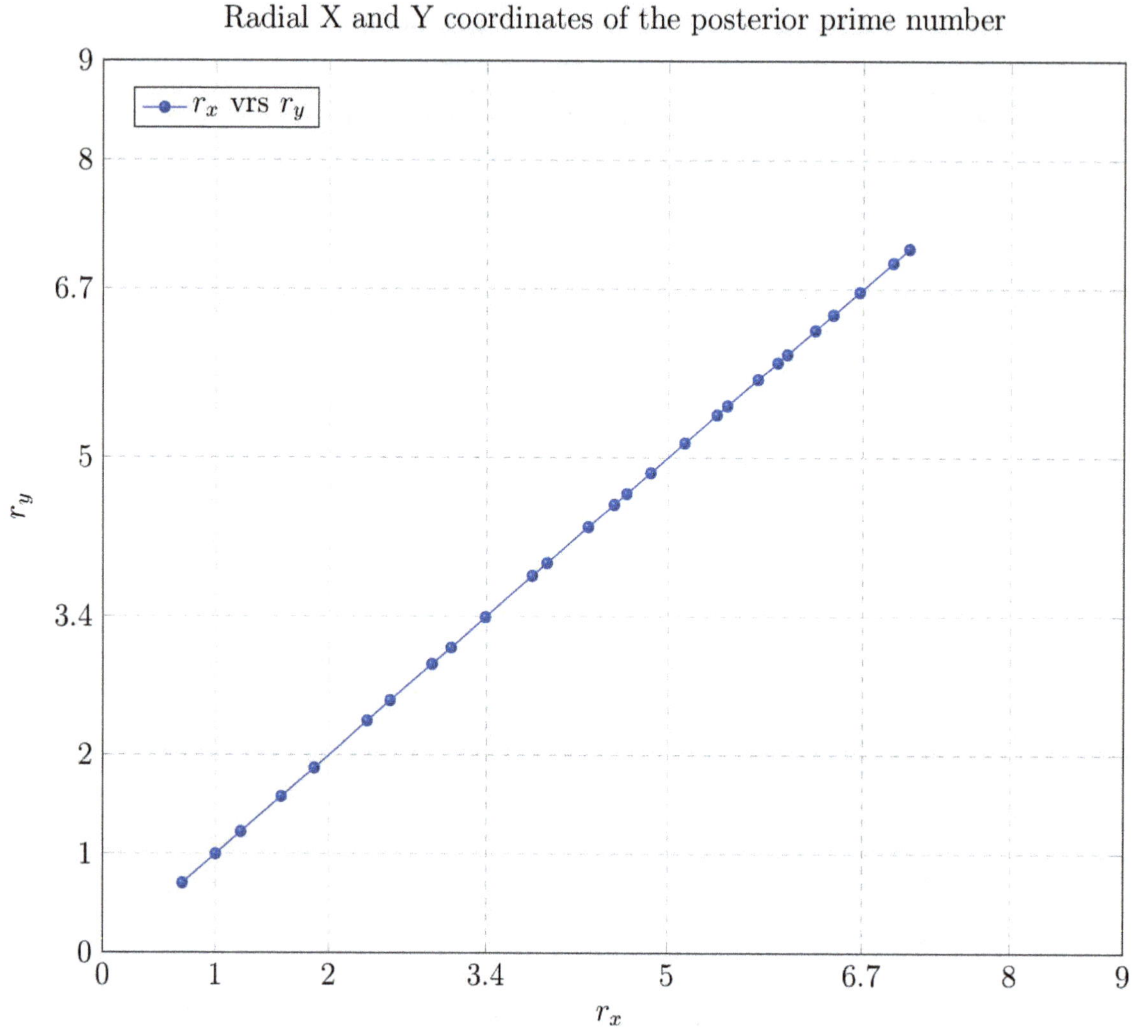

Figure 1.15: Radial X and Y coordinates of the posterior prime number, at 45 degrees, s=0.5, f=1.

Table 1.5: Prime Numbers Formula for +45 Degrees.

Facto	f=	0.7071	m=	1	Q=	1	k=	1	s=	1
Qty	**Posterior prime number partitions** p_1	**Previou Prime Number** r_1	**Modul** ΔL	**Modul** $\sum \Delta L$	**Posterior prime number** r_3	**Spiral Angle** θ **Rads**	**Spiral Angle** θ **Degs**	**Calculated Posterior Prime Number** r_3	**Coord X Real Posterior Prime Number** r_{x3}	**Coord Y Real Posterior Prime Number** r_{y3}
1	1	1	0	0	1	0.785398	45.000000	1.000000	0.707107	0.707107
2	0.750970	1	1	1	2	0.785398	45.000000	2.000000	1.414214	1.414214
3	1.497350	2	1	2	3	0.785398	45.000000	3.000000	2.121320	2.121320
4	2.147050	3	2	4	5	0.785398	45.000000	5.000000	3.535534	3.535534
5	2.752850	5	2	6	7	0.785398	45.000000	7.000000	4.949747	4.949747
6	3.319850	7	4	10	11	0.785398	45.000000	11.000000	7.778175	7.778175
7	3.765500	11	2	12	13	0.785398	45.000000	13.000000	9.192388	9.192388
8	4.079300	13	4	16	17	0.785398	45.000000	17.000000	12.020816	12.020816
9	4.353000	17	2	18	19	0.785398	45.000000	19.000000	13.435029	13.435029
10	4.570900	19	4	22	23	0.785398	45.000000	23.000000	16.263456	16.263456
11	4.875600	23	6	28	29	0.785398	45.000000	29.000000	20.506096	20.506096
12	5.091000	29	2	30	31	0.785398	45.000000	31.000000	21.920310	21.920310
13	5.266500	31	6	36	37	0.785398	45.000000	37.000000	26.162951	26.162951
14	5.468000	37	4	40	41	0.785398	45.000000	41.000000	28.991378	28.991378
15	5.576000	41	2	42	43	0.785398	45.000000	43.000000	30.405591	30.405591
16	5.675000	43	4	46	47	0.785398	45.000000	47.000000	33.234019	33.234019
17	5.826000	47	6	52	53	0.785398	45.000000	53.000000	37.476659	37.476659
18	5.990000	53	6	58	59	0.785398	45.000000	59.000000	41.719300	41.719300
19	6.090000	59	2	60	61	0.785398	45.000000	61.000000	43.133513	43.133513
20	6.183000	61	6	66	67	0.785398	45.000000	67.000000	47.376154	47.376154
21	6.294000	67	4	70	71	0.785398	45.000000	71.000000	50.204582	50.204582
22	6.355000	71	2	72	73	0.785398	45.000000	73.000000	51.618795	51.618795
23	6.432000	73	6	78	79	0.785398	45.000000	79.000000	55.861436	55.861436
24	6.525000	79	4	82	83	0.785398	45.000000	83.000000	58.689863	58.689863
25	6.610000	83	6	88	89	0.785398	45.000000	89.000000	62.932503	62.932503
26	6.724000	89	8	96	97	0.785398	45.000000	97.000000	68.589358	68.589358
27	6.815000	97	4	100	101	0.785398	45.000000	101.000000	71.417785	71.417785

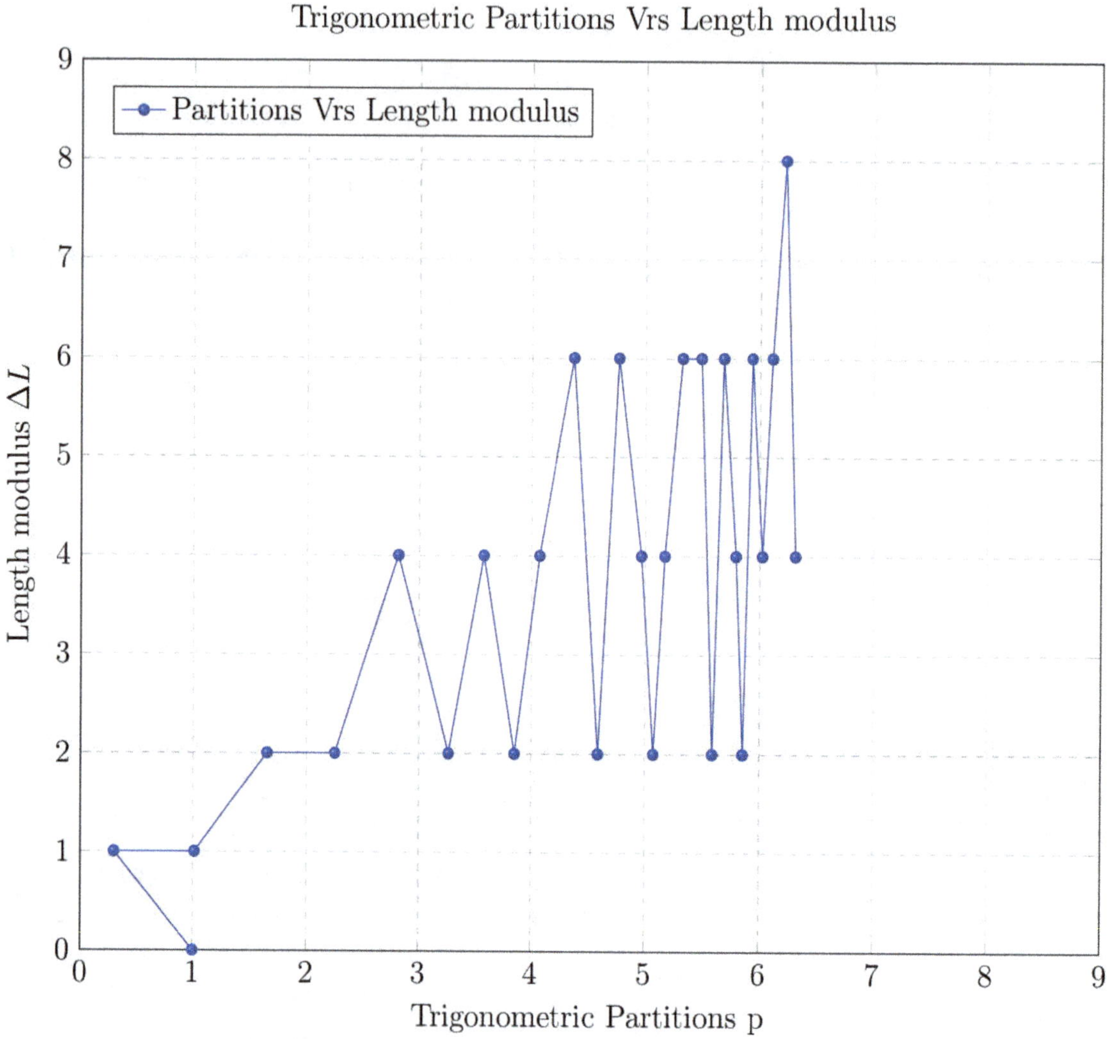

Figure 1.16: Trigonometric Partitions Vrs Length modulus, at 45 degrees, s=1,f=0.7071.

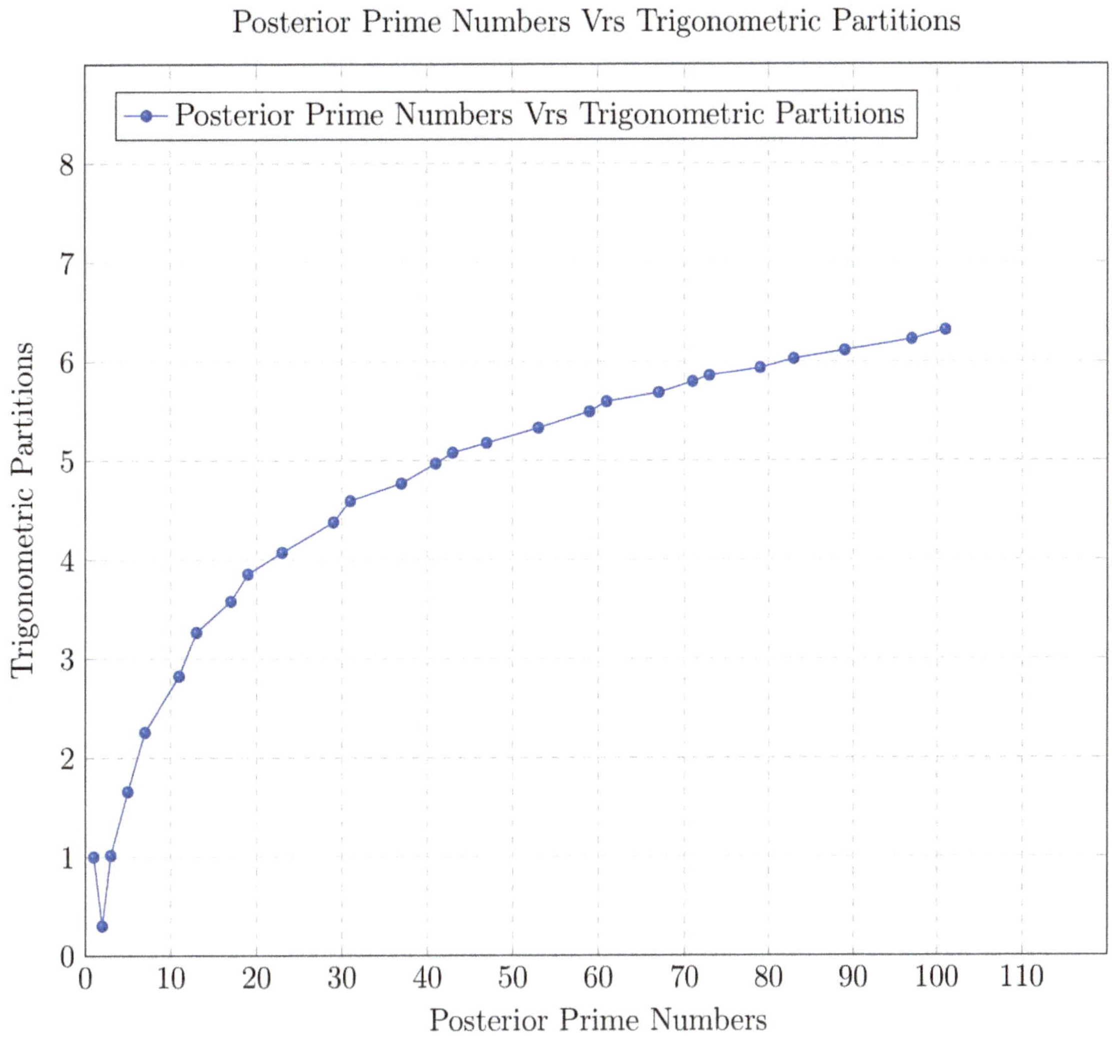

Figure 1.17: Posterior Prime Numbers Vrs Trigonometric Partitions, at 45 degrees, s=1,f=0.7071.

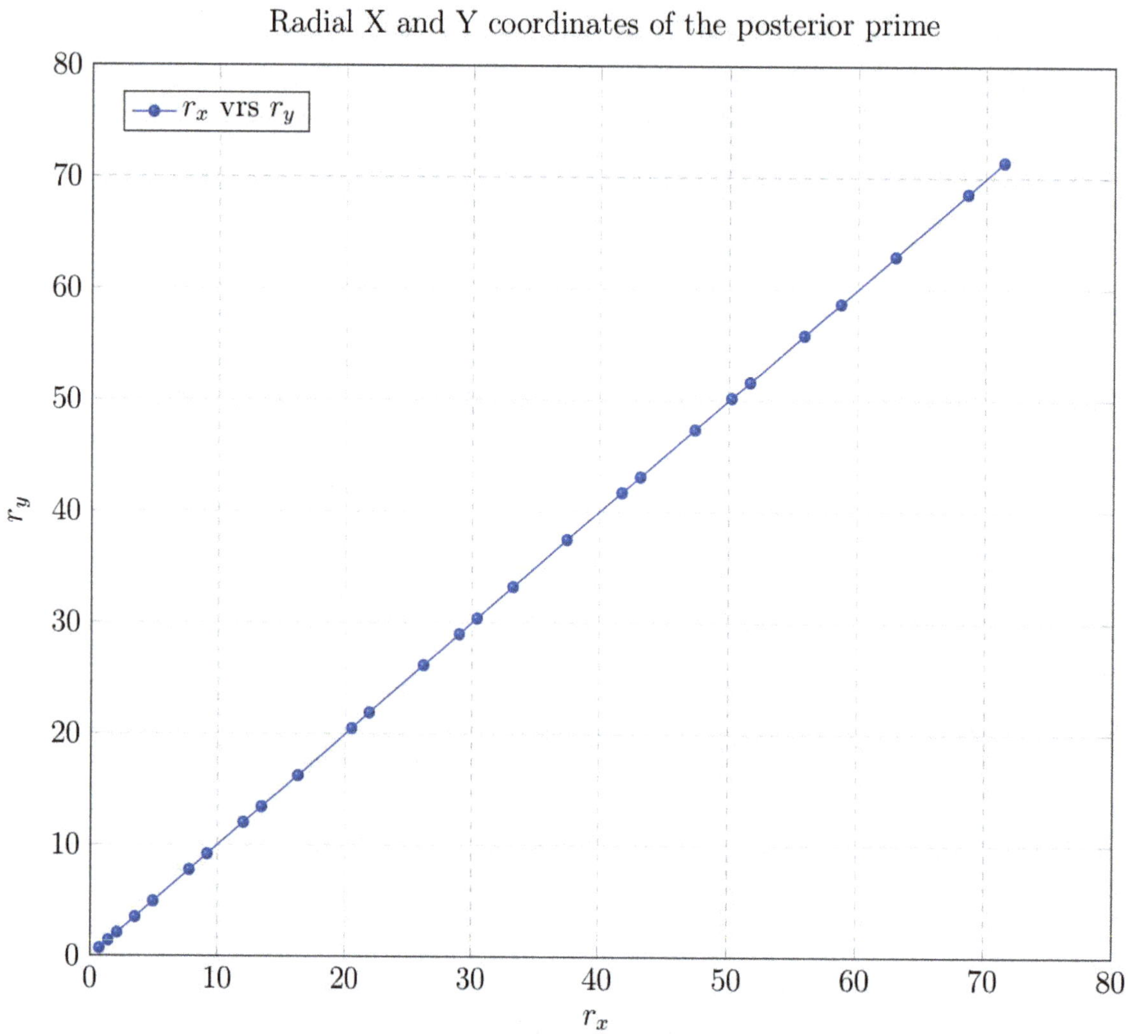

Figure 1.18: Radial X and Y coordinates of the posterior prime, at 45 degrees, s=1,f=0.7071.

Table 1.6: Prime Numbers Formula for +45 Degrees.

Facto	f=	1	m=	1	Q=	1	k=	1	s=	1
Qty	**Posterior prime number partitions p_1**	**Previou Prime Number r_1**	**Modul ΔL**	**Modul $\sum \Delta L$**	**Posterior prime number r_3**	**Spiral Angle θ Rads**	**Spiral Angle θ Degs**	**Calculated Posterior Prime Number r_3**	**Coord X Real Posterior Prime Number r_{x3}**	**Coord Y Real Posterior Prime Number r_{y3}**
1	1	1	0	0	1	0.785398	45.000000	1.000000	0.707107	0.707107
2	0.300840	1	1	1	2	0.785398	45.000000	2.000000	1.414214	1.414214
3	1.015725	2	1	2	3	0.785398	45.000000	3.000000	2.121321	2.121321
4	1.655540	3	2	4	5	0.785398	45.000000	5.000000	3.535534	3.535534
5	2.256250	5	2	6	7	0.785398	45.000000	7.000000	4.949747	4.949747
6	2.821500	7	4	10	11	0.785398	45.000000	11.000000	7.778174	7.778174
7	3.266300	11	2	12	13	0.785398	45.000000	13.000000	9.192388	9.192388
8	3.579800	13	4	16	17	0.785398	45.000000	17.000000	12.020815	12.020815
9	3.853100	17	2	18	19	0.785398	45.000000	19.000000	13.435029	13.435029
10	4.071000	19	4	22	23	0.785398	45.000000	23.000000	16.263456	16.263456
11	4.375900	23	6	28	29	0.785398	45.000000	29.000000	20.506097	20.506097
12	4.591000	29	2	30	31	0.785398	45.000000	31.000000	21.920310	21.920310
13	4.766900	31	6	36	37	0.785398	45.000000	37.000000	26.162951	26.162951
14	4.968500	37	4	40	41	0.785398	45.000000	41.000000	28.991378	28.991378
15	5.077000	41	2	42	43	0.785398	45.000000	43.000000	30.405592	30.405592
16	5.175000	43	4	46	47	0.785398	45.000000	47.000000	33.234018	33.234018
17	5.326000	47	6	52	53	0.785398	45.000000	53.000000	37.476659	37.476659
18	5.490000	53	6	58	59	0.785398	45.000000	59.000000	41.719300	41.719300
19	5.592000	59	2	60	61	0.785398	45.000000	61.000000	43.133514	43.133514
20	5.683000	61	6	66	67	0.785398	45.000000	67.000000	47.376154	47.376154
21	5.793000	67	4	70	71	0.785398	45.000000	71.000000	50.204582	50.204582
22	5.855000	71	2	72	73	0.785398	45.000000	73.000000	51.618795	51.618795
23	5.932000	73	6	78	79	0.785398	45.000000	79.000000	55.861436	55.861436
24	6.025000	79	4	82	83	0.785398	45.000000	83.000000	58.689863	58.689863
25	6.110000	83	6	88	89	0.785398	45.000000	89.000000	62.932503	62.932503
26	6.223000	89	8	96	97	0.785398	45.000000	97.000000	68.589358	68.589358
27	6.315000	97	4	100	101	0.785398	45.000000	101.000000	71.417785	71.417785

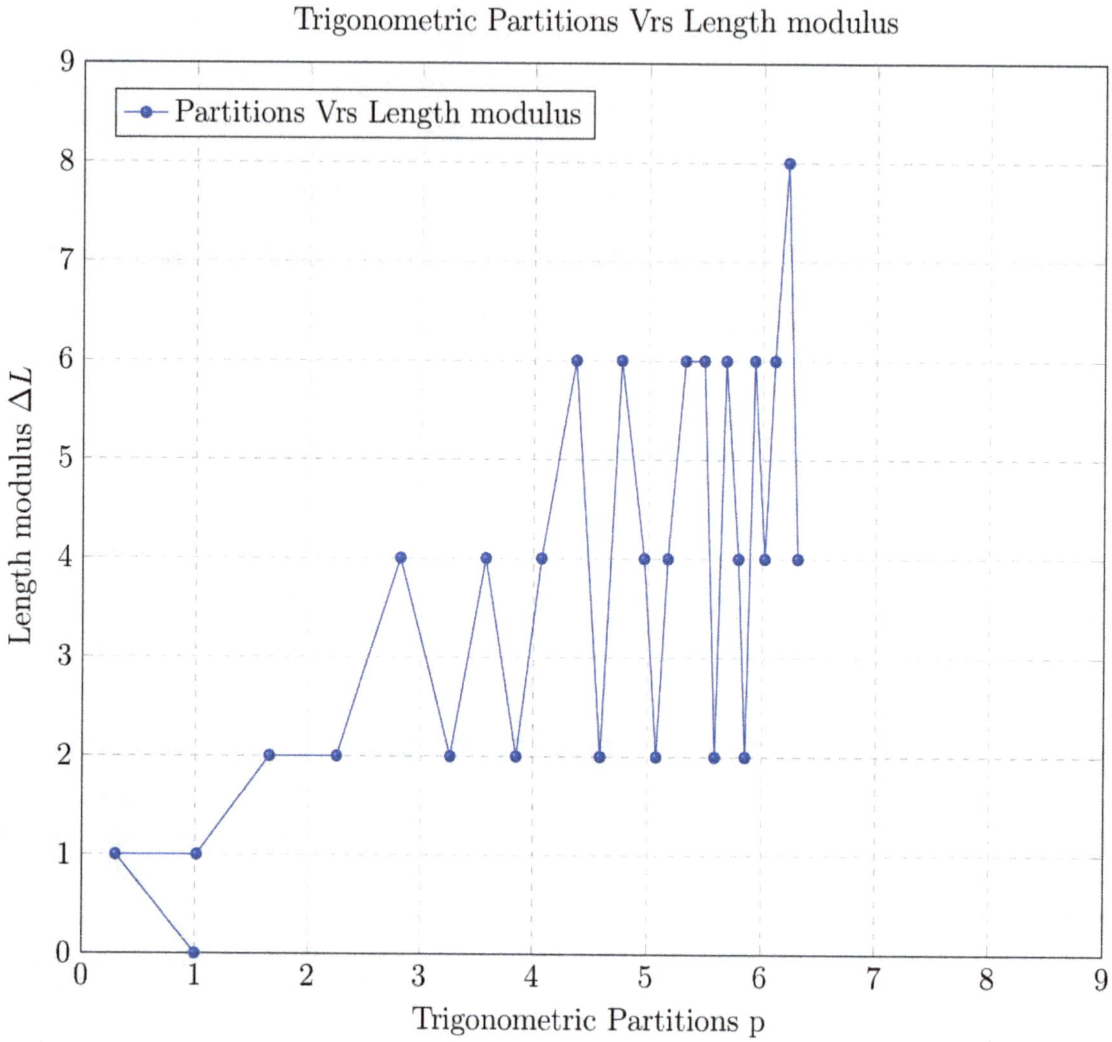

Figure 1.19: Trigonometric Partitions Vrs Length modulus, at 45 degrees, s=1,f=1.

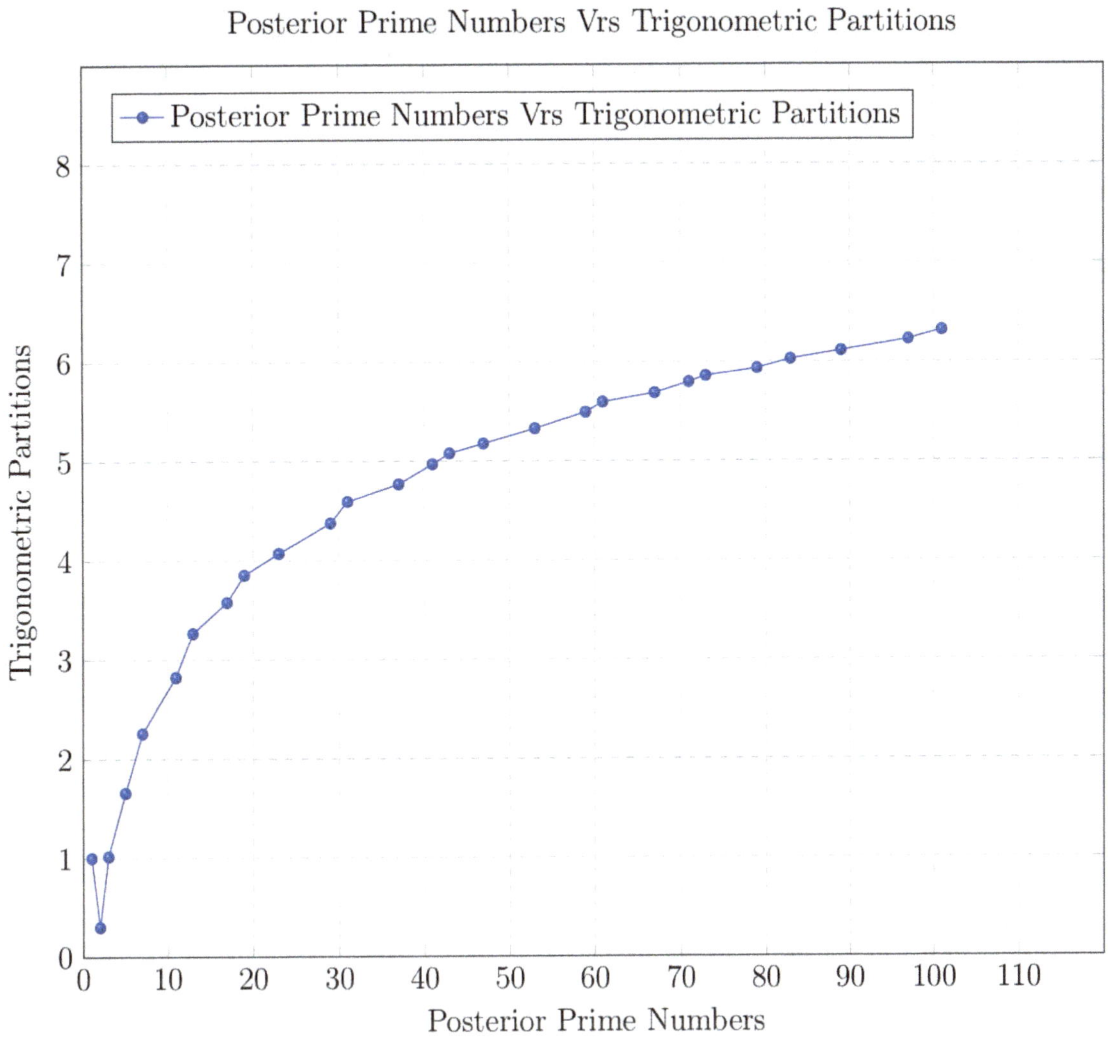

Figure 1.20: Posterior Prime Numbers Vrs Trigonometric Partitions, at 45 degrees, s=1,f=1.

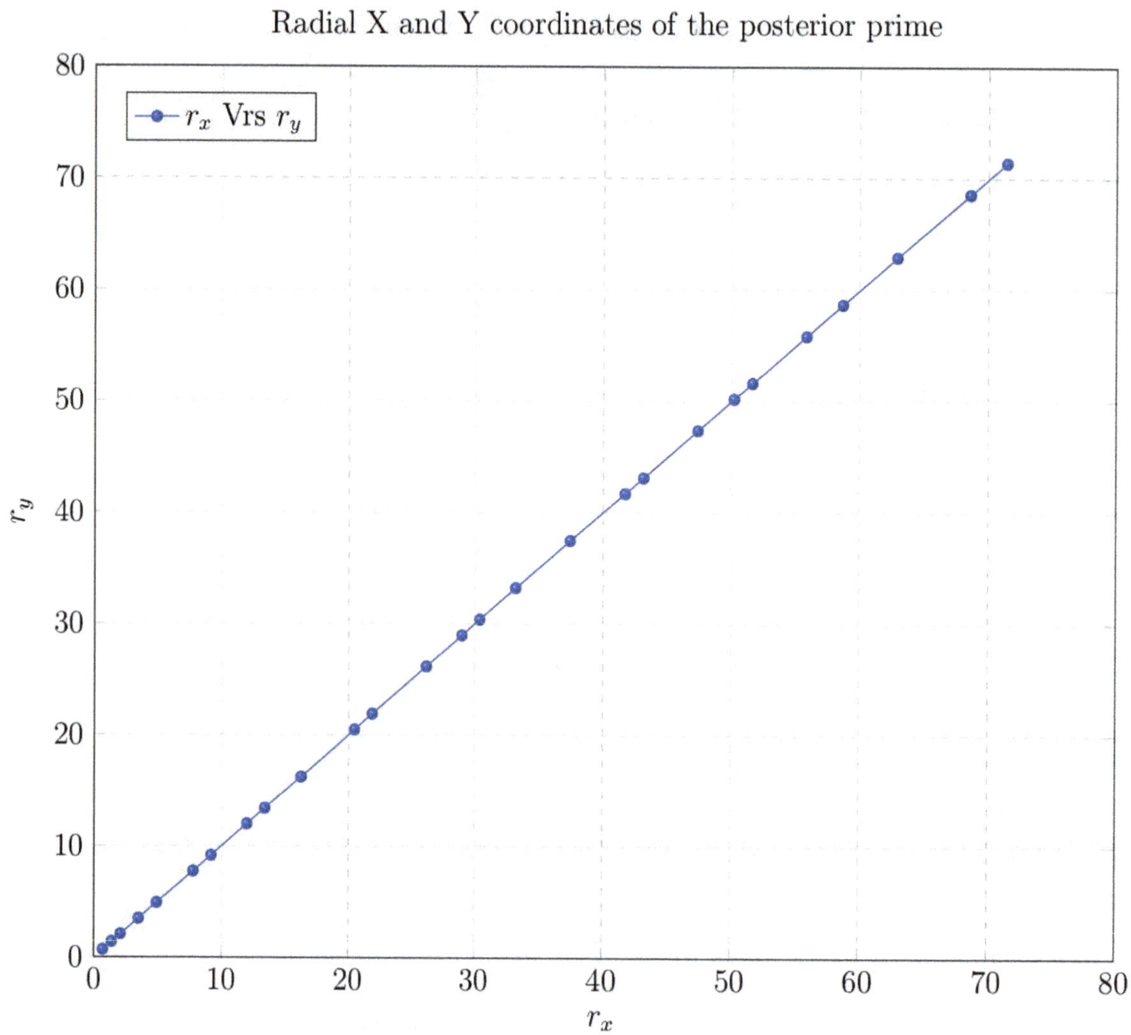

Figure 1.21: Radial X and Y coordinates of the posterior prime, at 45 degrees, s=1,f=1.

Table 1.7: Prime Numbers Formula for +45 Degrees.

Factc	f=	0.70711	m=	1	Q=	2	k=	1	s=	2
Qty	**Posterior prime number partitions** p_1	**Previous Prime Number** r_1	**Modulo** ΔL	**Modulo** $\sum \Delta L$	**Posterior prime number** r_3	**Spiral Angle** θ **Rads**	**Spiral Angle** θ **Degs**	**Calculated Posterior Prime Number** r_3	**Coord X Real Posterior Prime Number** r_{x3}	**Coord Y Real Posterior Prime Number** r_{y3}
1	1	1	0	0	1	0.785398	45	1.000000	0.707107	0.707107
2	1.251255	1	3	3	4	0.785398	45	4.000000	2.828427	2.828427
3	2.774500	4	5	8	9	0.785398	45	9.000000	6.363961	6.363961
4	4.092600	9	16	24	25	0.785398	45	25.000000	17.677669	17.677669
5	5.314100	25	24	48	49	0.785398	45	49.000000	34.648232	34.648232
6	6.451500	49	72	120	121	0.785398	45	121.000000	85.559920	85.559920
7	7.345000	121	48	168	169	0.785398	45	169.000000	119.501046	119.501046
8	7.973000	169	120	288	289	0.785398	45	289.000000	204.353860	204.353860
9	8.520000	289	72	360	361	0.785398	45	361.000000	255.265548	255.265548
10	8.955000	361	168	528	529	0.785398	45	529.000000	374.059487	374.059487
11	9.565000	529	312	840	841	0.785398	45	841.000000	594.676803	594.676803
12	9.989000	841	120	960	961	0.785398	45	961.000000	679.529616	679.529616
13	10.350000	961	408	1368	1369	0.785398	45	1369.000000	968.029184	968.029184
14	10.750000	1369	312	1680	1681	0.785398	45	1681.000000	1188.646499	1188.646499
15	10.950000	1681	168	1848	1849	0.785398	45	1849.000000	1307.440438	1307.440438
16	11.150000	1849	360	2208	2209	0.785398	45	2209.000000	1561.998879	1561.998879
17	11.460000	2209	600	2808	2809	0.785398	45	2809.000000	1986.262948	1986.262948
18	11.800000	2809	672	3480	3481	0.785398	45	3481.000000	2461.438705	2461.438705
19	12.000000	3481	240	3720	3721	0.785398	45	3721.000000	2631.144333	2631.144333
20	12.200000	3721	768	4488	4489	0.785398	45	4489.000000	3174.202341	3174.202341
21	12.350000	4489	552	5040	5041	0.785398	45	5041.000000	3564.525284	3564.525284
22	12.650000	5041	288	5328	5329	0.785398	45	5329.000000	3768.172037	3768.172037
23	12.700000	5329	912	6240	6241	0.785398	45	6241.000000	4413.053422	4413.053422
24	12.900000	6241	648	6888	6889	0.785398	45	6889.000000	4871.258616	4871.258616
25	13.000000	6889	1032	7920	7921	0.785398	45	7921.000000	5600.992814	5600.992814
26	13.200000	7921	1488	9408	9409	0.785398	45	9409.000000	6653.167704	6653.167704
27	13.500000	9409	792	10200	10201	0.785398	45	10201.000000	7213.196275	7213.196275

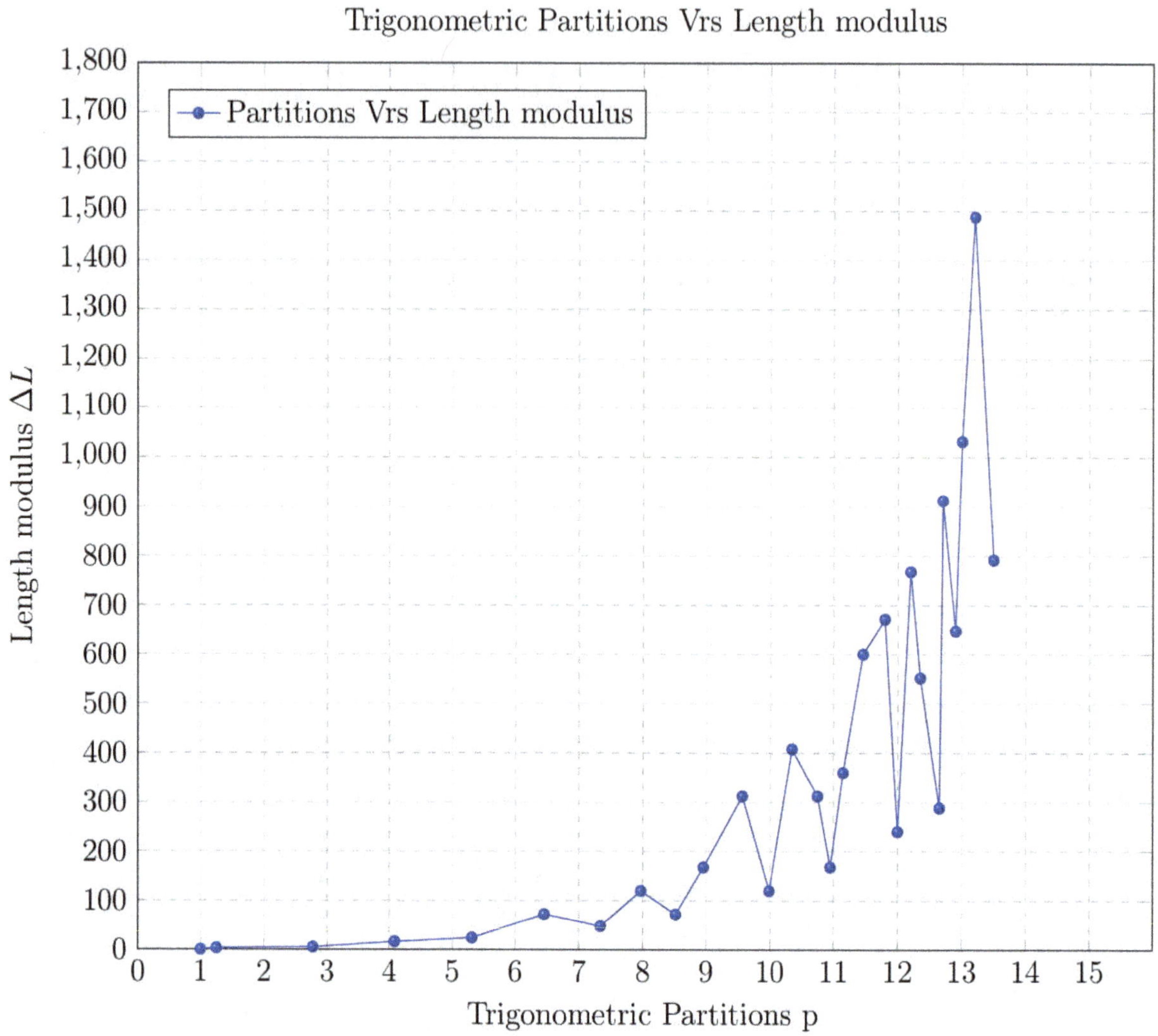

Figure 1.22: Trigonometric Partitions Vrs Length modulus, at 45 degrees, s=2,f=0.70711.

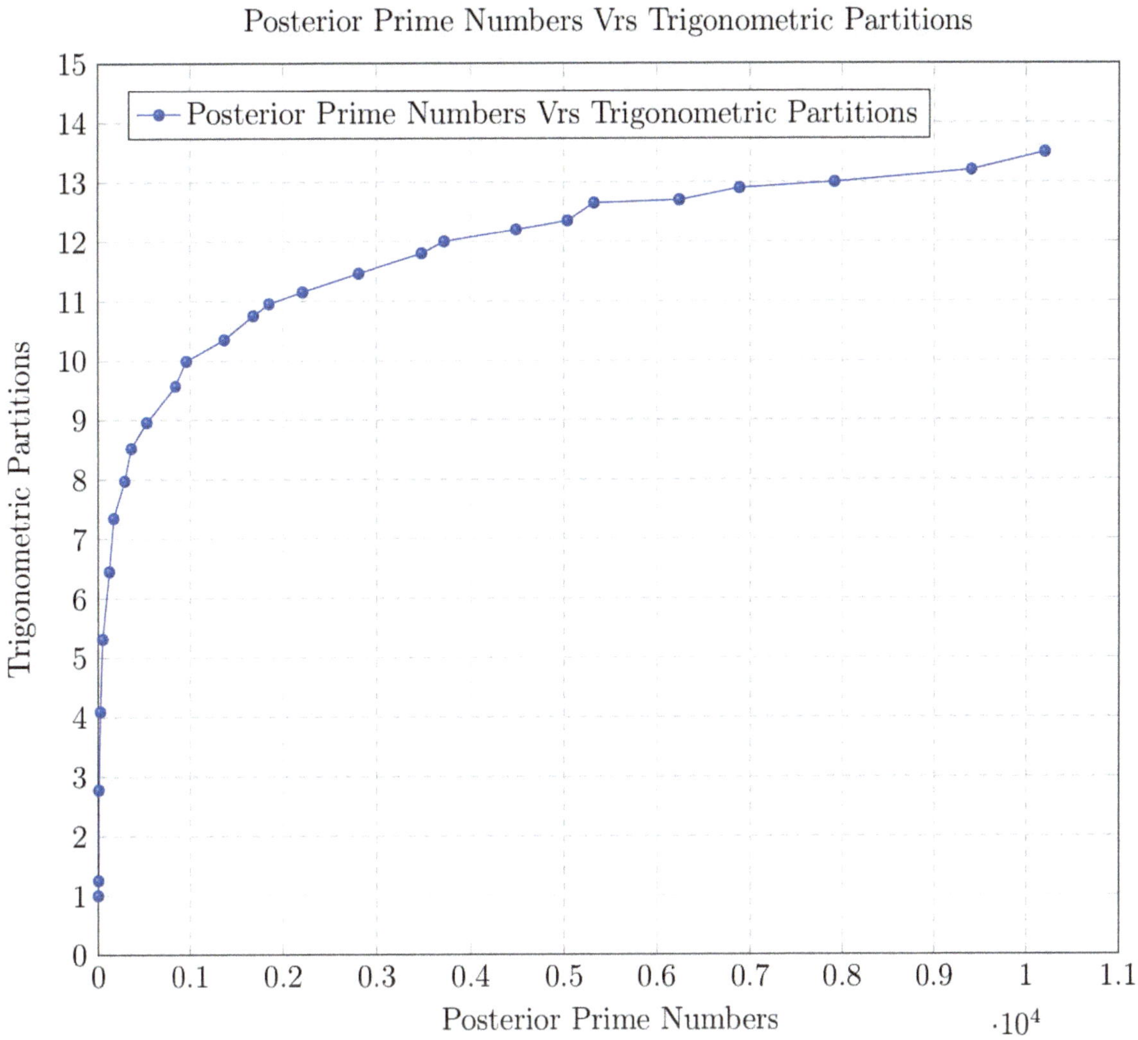

Figure 1.23: Posterior Prime Numbers Vrs Trigonometric Partitions, at 45 degrees, s=2,f=0.70711.

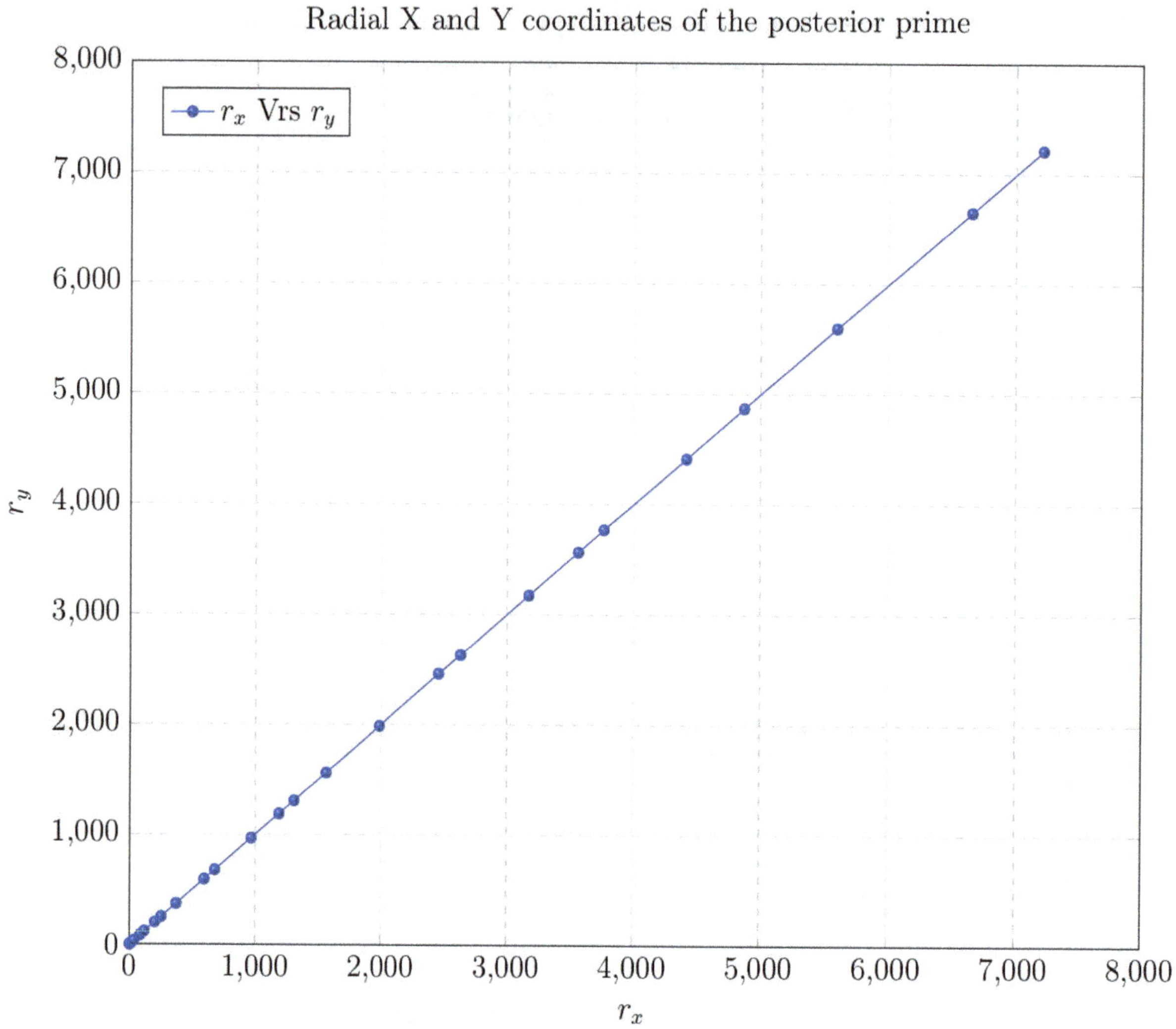

Figure 1.24: Radial X and Y coordinates of the posterior prime, at 45 degrees, s=2, f=0.70711.

With the initial angle constant.;$\theta = 60 \quad = \pi/3$;The Trigonometric Partitions are culated by trial and error; $r_3(p) = \sqrt{[\Delta L * e^{[2^{(p-1)} * \sin(\pi * k/(3 * 2^{(p+1)})) - \pi \ k/3]} + \quad (r_1^2 + f^2)]^2 - f^2}$.

Table 1.8: Prime Numbers Formula for $\theta_0 = +60$ Degrees.

Factc	f=	0.86603	m=	0	Q=	2	k=	1	s=	0
Qty	Posterior prime number partitions p_1	Previous Prime Number r_1	Modul ΔL	Modul $\sum \Delta L$	Posterior prime number r_3	Spiral Angle θ Rads	Spiral Angle θ Degs	Calculated Posterior Prime Number r_3	Coord X Real Posterior Prime Number r_{x3}	Coord Y Real Posterior Prime Number r_{y3}
1	1	1	0	0	1	1.047198	60	1	0.500000	0.866025
2	1.1194600	1	0	0	1	1.047198	60	1	0.500000	0.866025
3	1.8410200	1	0	0	1	1.047198	60	1	0.500000	0.866025
4	2.4831200	1	0	0	1	1.047198	60	1	0.500000	0.866025
5	3.0853600	1	0	0	1	1.047198	60	1	0.500000	0.866025
6	3.6511000	1	0	0	1	1.047198	60	1	0.500000	0.866025
7	4.0962000	1	0	0	1	1.047198	60	1	0.500000	0.866025
8	4.4097000	1	0	0	1	1.047198	60	1	0.500000	0.866025
9	4.6830000	1	0	0	1	1.047198	60	1	0.500000	0.866025
10	4.9010000	1	0	0	1	1.047198	60	1	0.500000	0.866025
11	5.2059000	1	0	0	1	1.047198	60	1	0.500000	0.866025
12	5.4210000	1	0	0	1	1.047198	60	1	0.500000	0.866025
13	5.59700000	1	0	0	1	1.047198	60	1	0.500000	0.866025
14	5.79850000	1	0	0	1	1.047198	60	1	0.500000	0.866025
15	5.90700000	1	0	0	1	1.047198	60	1	0.500000	0.866025
16	6.00500000	1	0	0	1	1.047198	60	1	0.500000	0.866025
17	6.15600000	1	0	0	1	1.047198	60	1	0.500000	0.866025
18	6.32000000	1	0	0	1	1.047198	60	1	0.500000	0.866025
19	6.42000000	1	0	0	1	1.047198	60	1	0.500000	0.866025
20	6.51300000	1	0	0	1	1.047198	60	1	0.500000	0.866025
21	6.62300000	1	0	0	1	1.047198	60	1	0.500000	0.866025
22	6.68500000	1	0	0	1	1.047198	60	1	0.500000	0.866025
23	6.76100000	1	0	0	1	1.047198	60	1	0.500000	0.866025
24	6.85500000	1	0	0	1	1.047198	60	1	0.500000	0.866025
25	6.94000000	1	0	0	1	1.047198	60	1	0.500000	0.866025
26	7.05200000	1	0	0	1	1.047198	60	1	0.500000	0.866025
27	7.14500000	1	0	0	1	1.047198	60	1	0.500000	0.866025

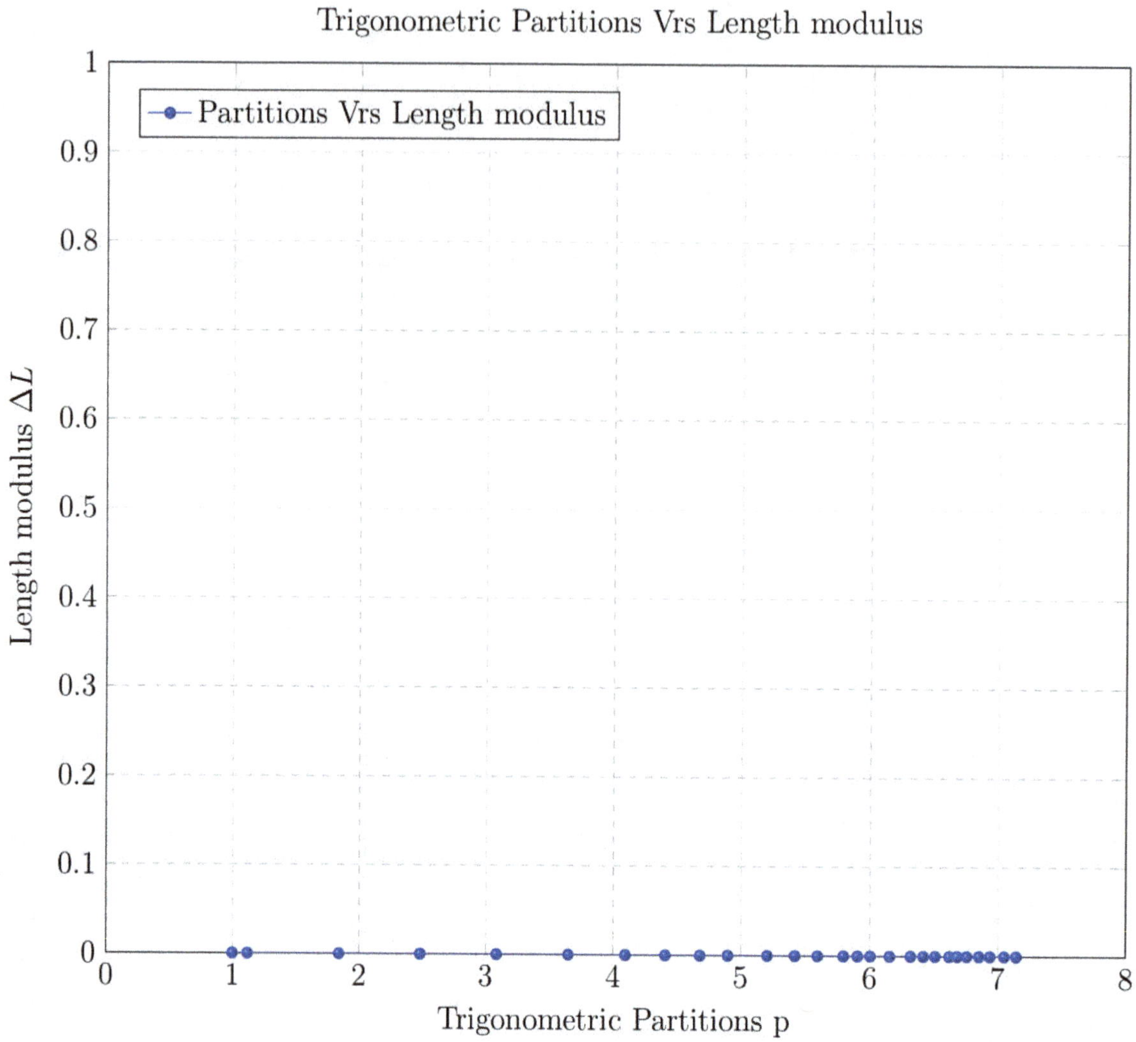

Figure 1.25: Trigonometric Partitions Vrs Length modulus, at 60 degrees, s=0,f=0.86603.

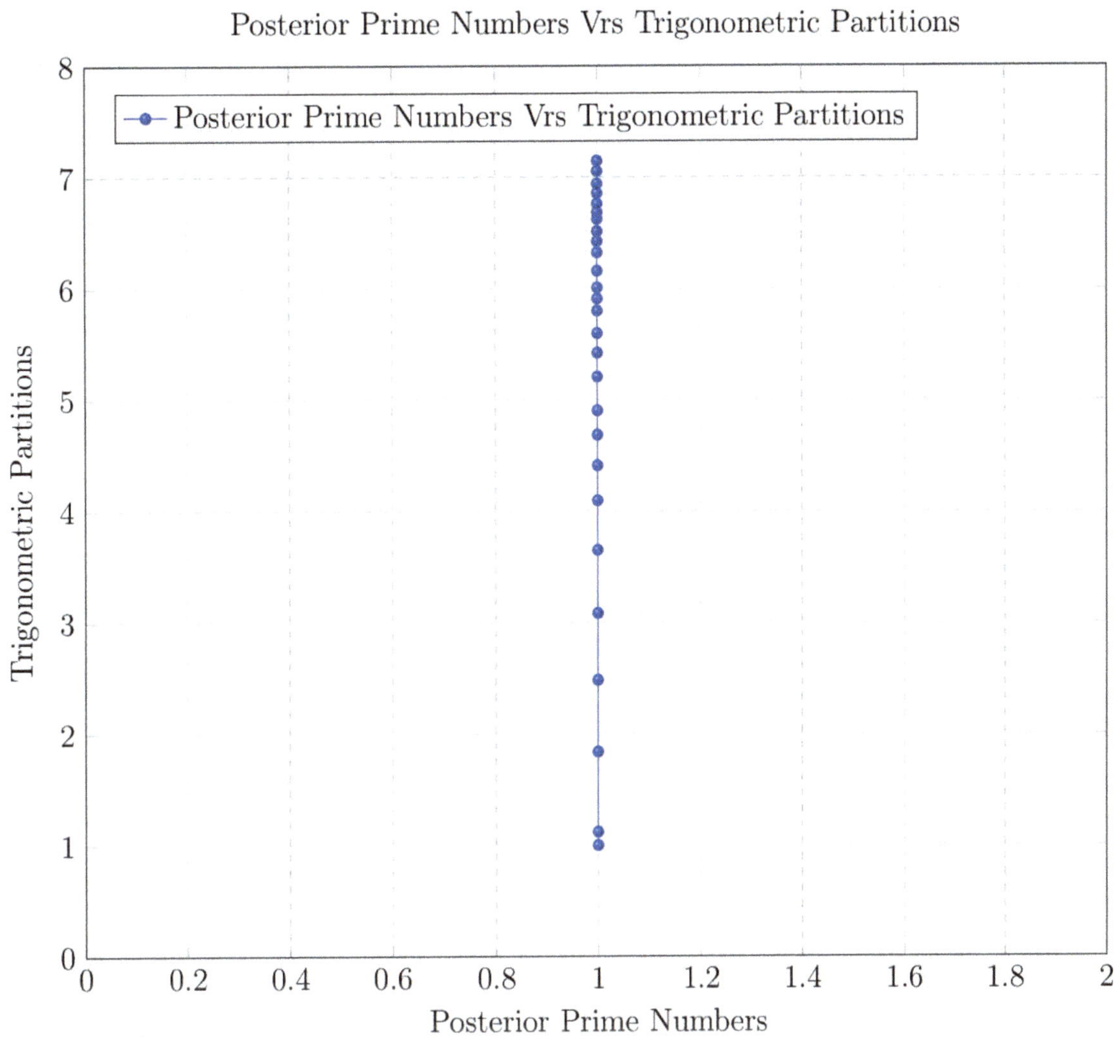

Figure 1.26: Posterior Prime Numbers Vrs Trigonometric Partitions, at 60 degrees, s=0,f=0.86603.

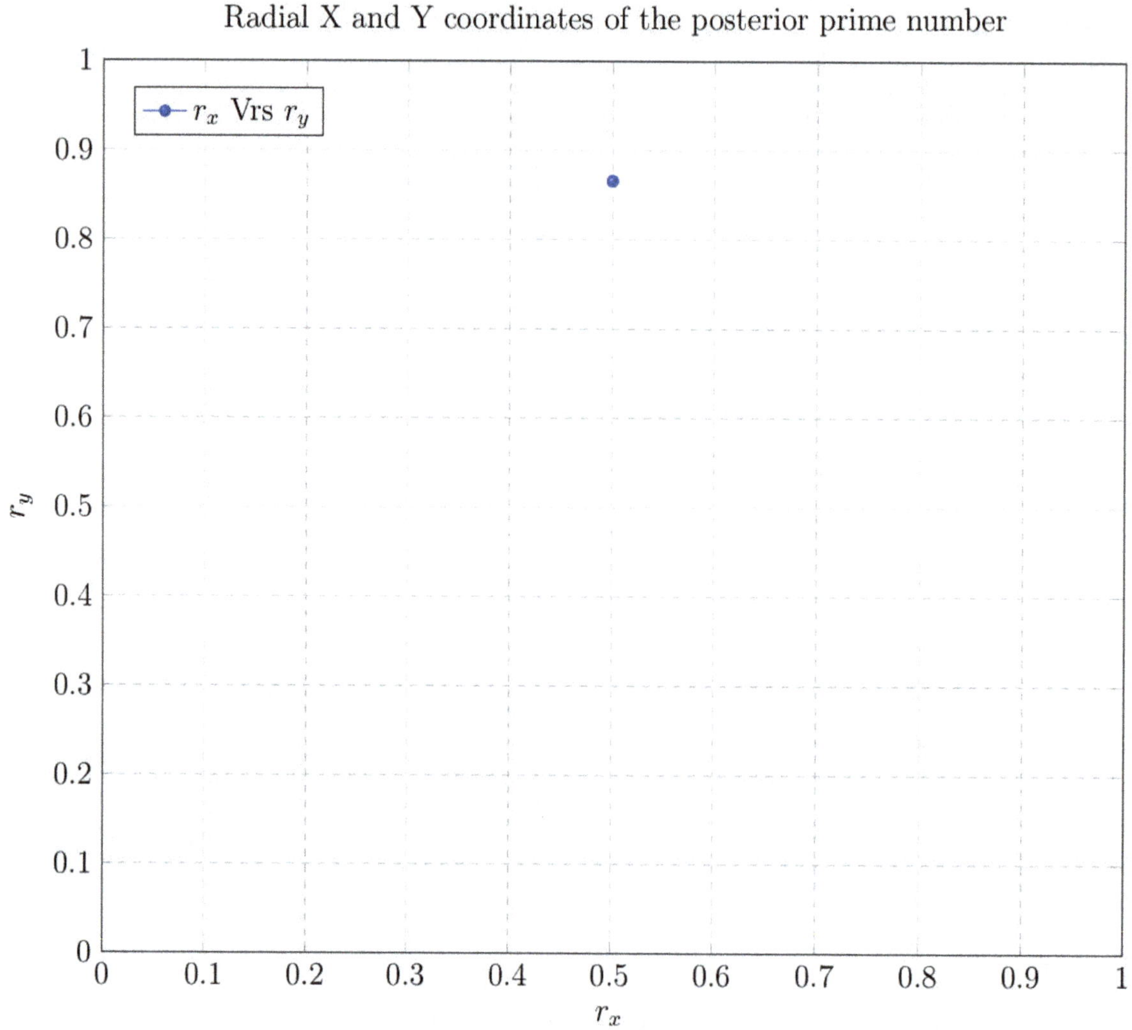

Figure 1.27: Radial X and Y coordinates of the posterior prime number, at 60 degrees, s=0,f=0.86603.

Table 1.9: Prime Numbers Formula for +60 Degrees.

Facto	f=	0.86603	m=	1	Q=	1	k=	1	s=	1
Qty	Posterior prime number partitions p_1	Previous Prime Number r_1	Module ΔL	Module $\sum \Delta L$	Posterior prime number r_3	Spiral Angle θ Rads	Spiral Angle θ Degs	Calculated Posterior Prime Number r_3	Coord X Real Posterior Prime Number r_{x3}	Coord Y Real Posterior Prime Number r_{y3}
1	1	1	0	0	1	1.047198	60	1.000000	0.500000	0.866025
2	1.1194600	1	1	1	2	1.047198	60	2.000000	1.000000	1.732051
3	1.8410200	2	1	2	3	1.047198	60	3.000000	1.500000	2.598076
4	2.4831200	3	2	4	5	1.047198	60	5.000000	2.500000	4.330127
5	3.0853600	5	2	6	7	1.047198	60	7.000000	3.500000	6.062178
6	3.6511000	7	4	10	11	1.047198	60	11.000000	5.500000	9.526280
7	4.0962000	11	2	12	13	1.047198	60	13.000000	6.500000	11.258331
8	4.4097000	13	4	16	17	1.047198	60	17.000000	8.500000	14.722432
9	4.6830000	17	2	18	19	1.047198	60	19.000000	9.500000	16.454482
10	4.9010000	19	4	22	23	1.047198	60	23.000000	11.500000	19.918584
11	5.2059000	23	6	28	29	1.047198	60	29.000000	14.500000	25.114737
12	5.4210000	29	2	30	31	1.047198	60	31.000000	15.500000	26.846787
13	5.59700000	31	6	36	37	1.047198	60	37.000000	18.500000	32.042940
14	5.79850000	37	4	40	41	1.047198	60	41.000000	20.500000	35.507042
15	5.90700000	41	2	42	43	1.047198	60	43.000000	21.500000	37.239092
16	6.00500000	43	4	46	47	1.047198	60	47.000000	23.500000	40.703194
17	6.15600000	47	6	52	53	1.047198	60	53.000000	26.500000	45.899346
18	6.32000000	53	6	58	59	1.047198	60	59.000000	29.500000	51.095499
19	6.42000000	59	2	60	61	1.047198	60	61.000000	30.500000	52.827549
20	6.51300000	61	6	66	67	1.047198	60	67.000000	33.500000	58.023702
21	6.62300000	67	4	70	71	1.047198	60	71.000000	35.500000	61.487804
22	6.68500000	71	2	72	73	1.047198	60	73.000000	36.500000	63.219855
23	6.76100000	73	6	78	79	1.047198	60	79.000000	39.500000	68.416007
24	6.85500000	79	4	82	83	1.047198	60	83.000000	41.500000	71.880109
25	6.94000000	83	6	88	89	1.047198	60	89.000000	44.500000	77.076261
26	7.05200000	89	8	96	97	1.047198	60	97.000000	48.500000	84.004464
27	7.14500000	97	4	100	101	1.047198	60	101.000000	50.500000	87.468566

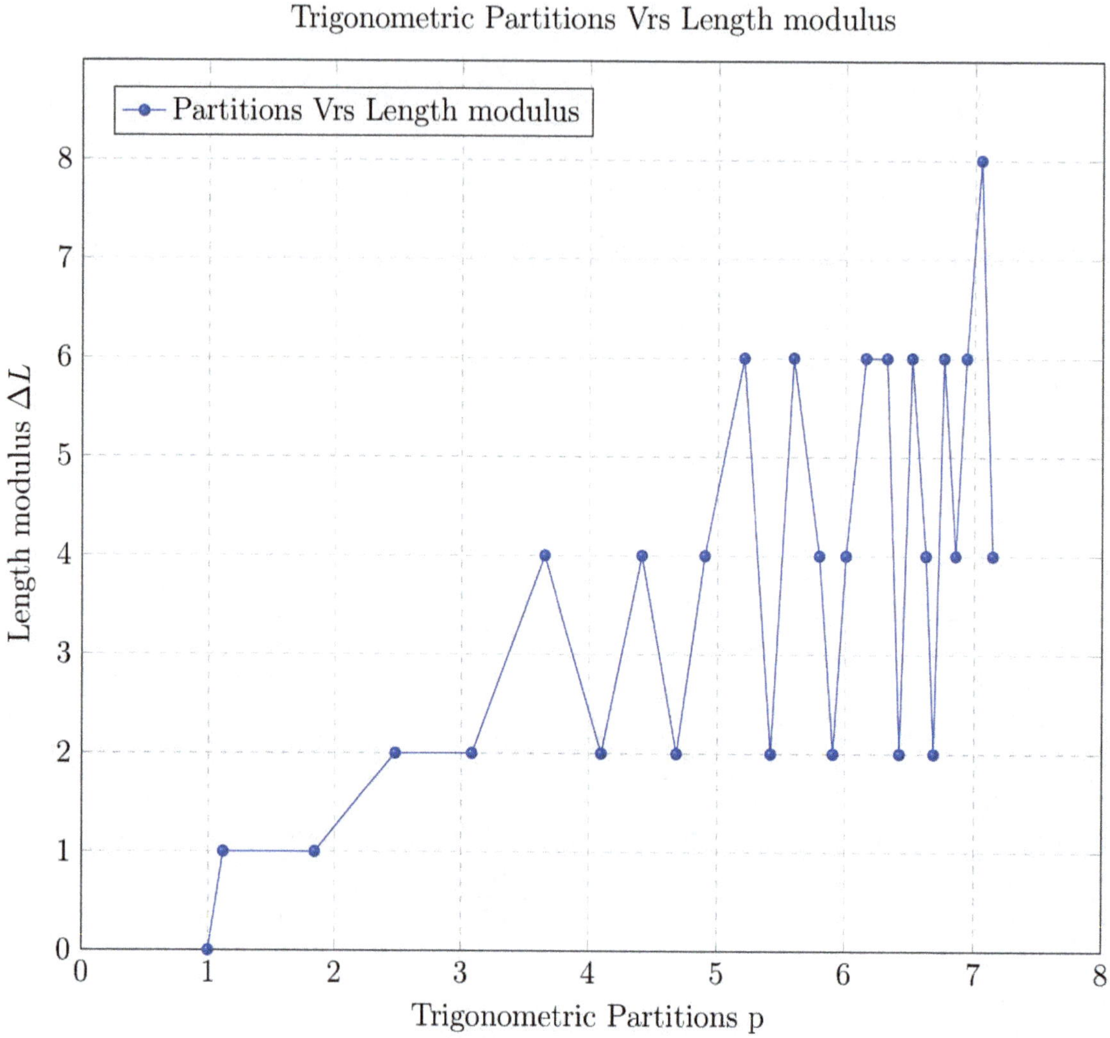

Figure 1.28: Trigonometric Partitions Vrs Length modulus, at 60 degrees, s=1,f=0.86603.

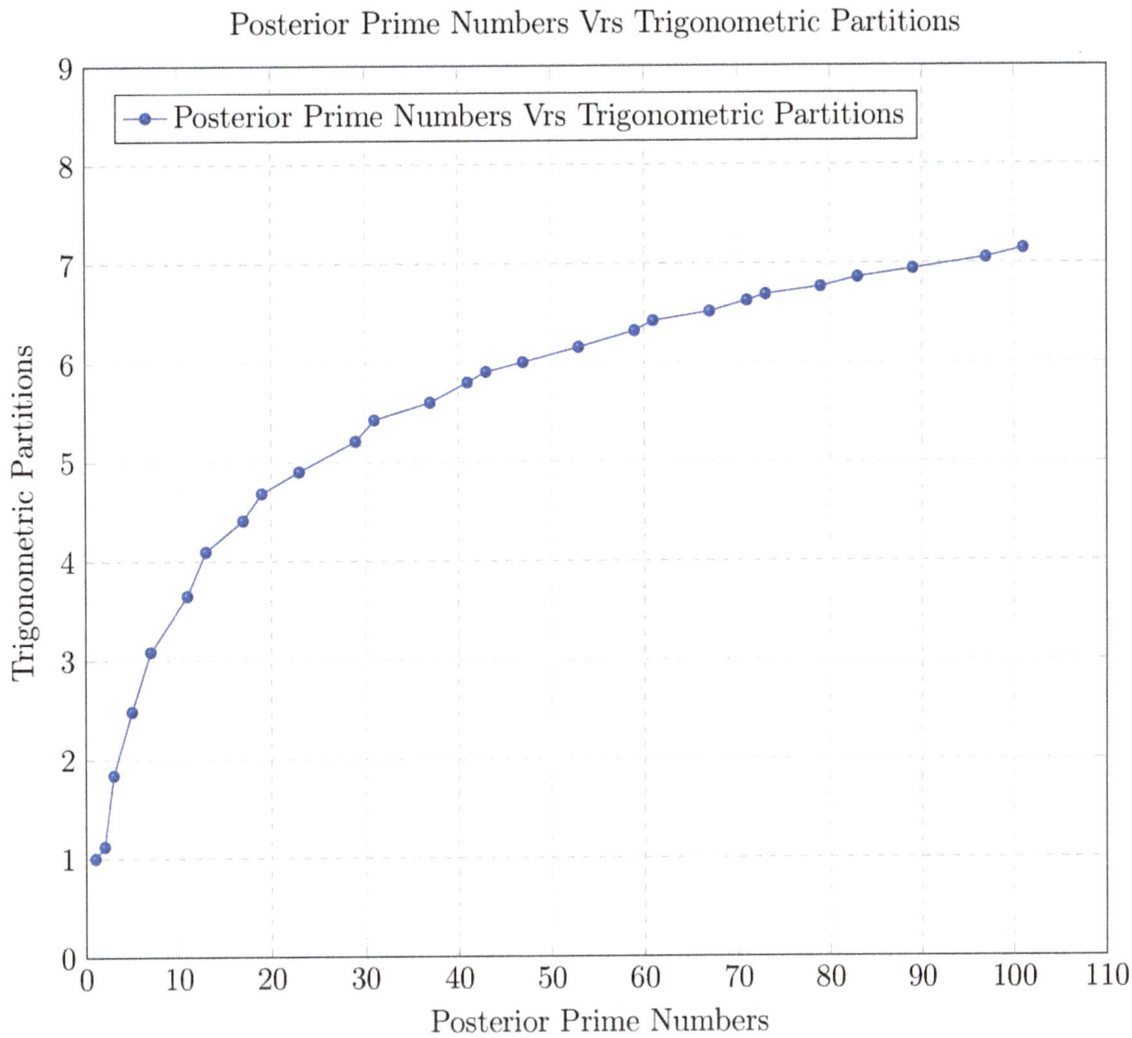

Figure 1.29: Posterior Prime Numbers Vrs Trigonometric Partitions, at 60 degrees, s=1,f=0.86603.

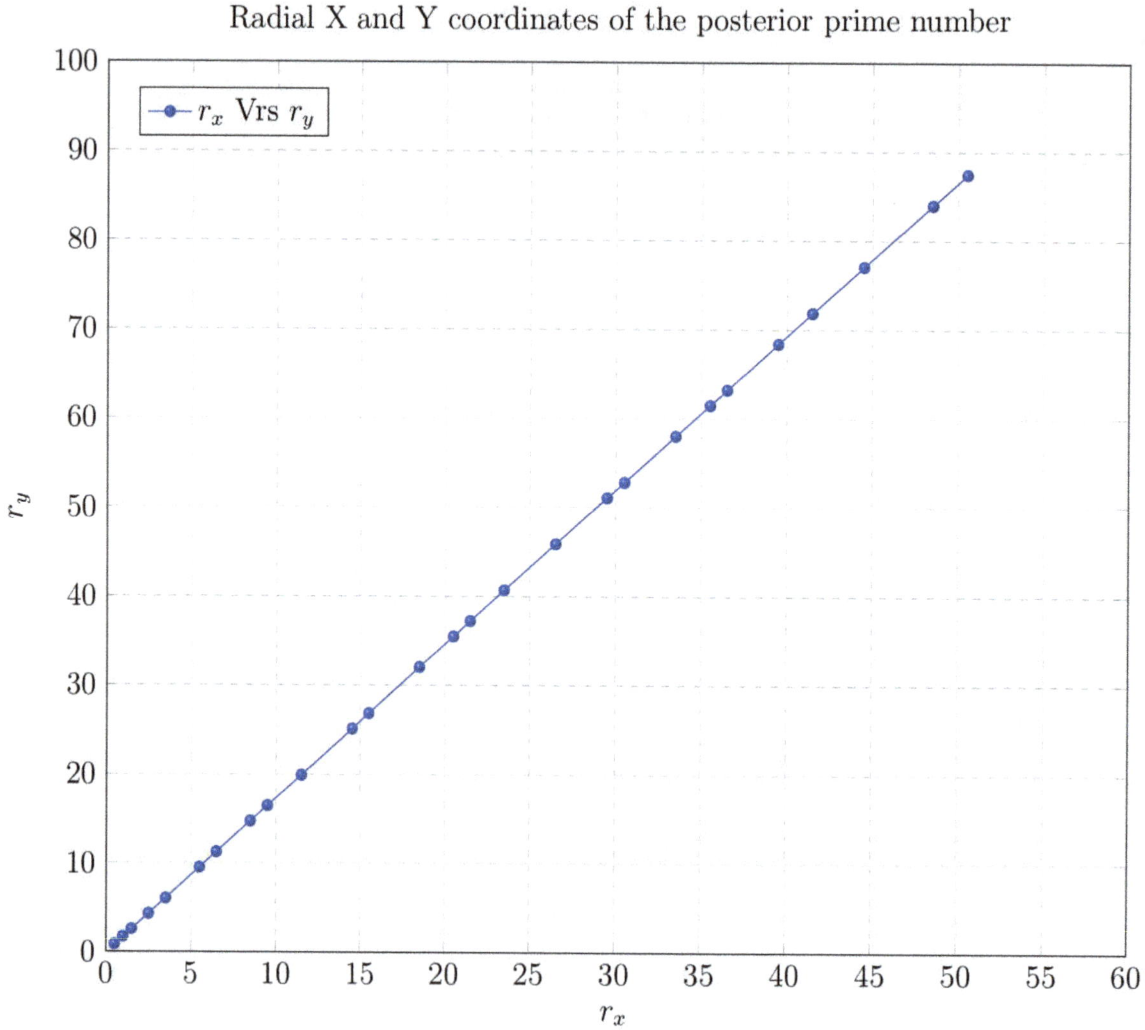

Figure 1.30: Radial X and Y coordinates of the posterior prime number, at 60 degrees, s=1,f=0.86603.

Table 1.10: Prime Numbers Formula for +60 Degrees.

Facto	f=	1	m=	1	Q=	1	k=	1	s=	1
Qty	**Posterior prime number partitions** p_1	**Previous Prime Number** r_1	**Module** ΔL	**Module** $\sum \Delta L$	**Posterior prime number** r_3	**Spiral Angle** θ **Rads**	**Spiral Angle** θ **Degs**	**Calculated Posterior Prime Number** r_3	**Coord X Real Posterior Prime Number** r_{x3}	**Coord Y Real Posterior Prime Number** r_{y3}
1	1	1	0	0	1	1.047198	60	1.000000	0.500000	0.866025
2	0.9390500	1	1	1	2	1.047198	60	2.000000	1.000000	1.732051
3	1.6438500	2	1	2	3	1.047198	60	3.000000	1.500000	2.598076
4	2.2803500	3	2	4	5	1.047198	60	5.000000	2.500000	4.330127
5	2.8798000	5	2	6	7	1.047198	60	7.000000	3.500000	6.062178
6	3.4445000	7	4	10	11	1.047198	60	11.000000	5.500000	9.526279
7	3.8891000	11	2	12	13	1.047198	60	13.000000	6.500000	11.258330
8	4.2025000	13	4	16	17	1.047198	60	17.000000	8.500000	14.722432
9	4.4759000	17	2	18	19	1.047198	60	19.000000	9.500000	16.454483
10	4.6936000	19	4	22	23	1.047198	60	23.000000	11.500000	19.918584
11	4.9985000	23	6	28	29	1.047198	60	29.000000	14.500000	25.114737
12	5.2140000	29	2	30	31	1.047198	60	31.000000	15.500000	26.846788
13	5.38950000	31	6	36	37	1.047198	60	37.000000	18.500000	32.042940
14	5.40000000	37	4	40	41	1.047198	60	40.999601	20.499800	35.506696
15	5.50000000	41	2	42	43	1.047198	60	42.999820	21.499910	37.238936
16	5.79800000	43	4	46	47	1.047198	60	47.000000	23.500000	40.703194
17	5.94850000	47	6	52	53	1.047198	60	53.000000	26.500000	45.899346
18	6.11250000	53	6	58	59	1.047198	60	59.000000	29.500000	51.095499
19	6.21500000	59	2	60	61	1.047198	60	61.000000	30.500000	52.827550
20	6.30600000	61	6	66	67	1.047198	60	67.000000	33.500000	58.023702
21	6.41500000	67	4	70	71	1.047198	60	71.000000	35.500000	61.487804
22	6.47600000	71	2	72	73	1.047198	60	73.000000	36.500000	63.219854
23	6.55350000	73	6	78	79	1.047198	60	79.000000	39.500000	68.416006
24	6.64600000	79	4	82	83	1.047198	60	83.000000	41.500000	71.880108
25	6.73300000	83	6	88	89	1.047198	60	89.000000	44.500000	77.076261
26	6.84500000	89	8	96	97	1.047198	60	97.000000	48.500000	84.004464
27	6.93500000	97	4	100	101	1.047198	60	101.000000	50.500000	87.468565

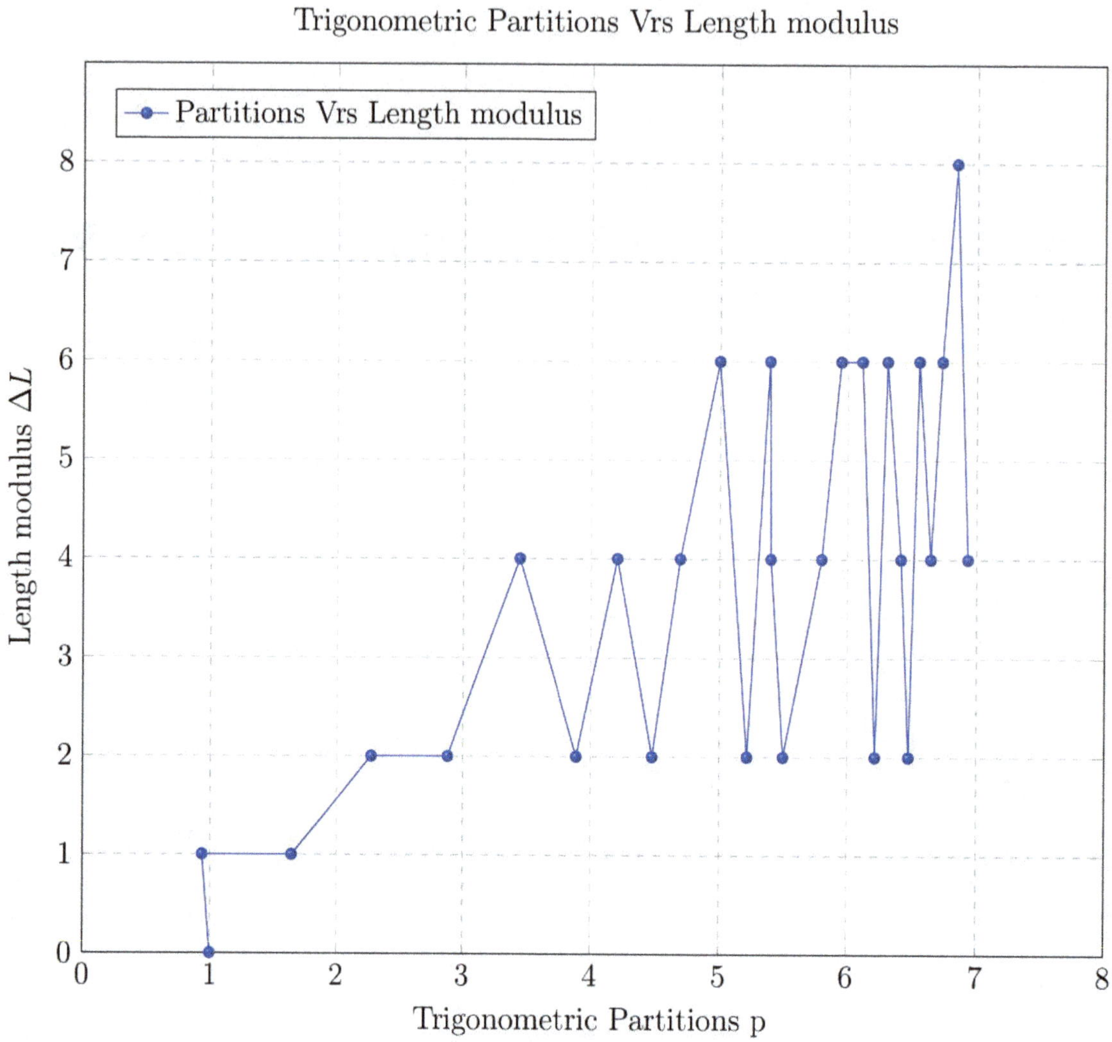

Figure 1.31: Trigonometric Partitions Vrs Length modulus, at 60 degrees, s=1,f=1.

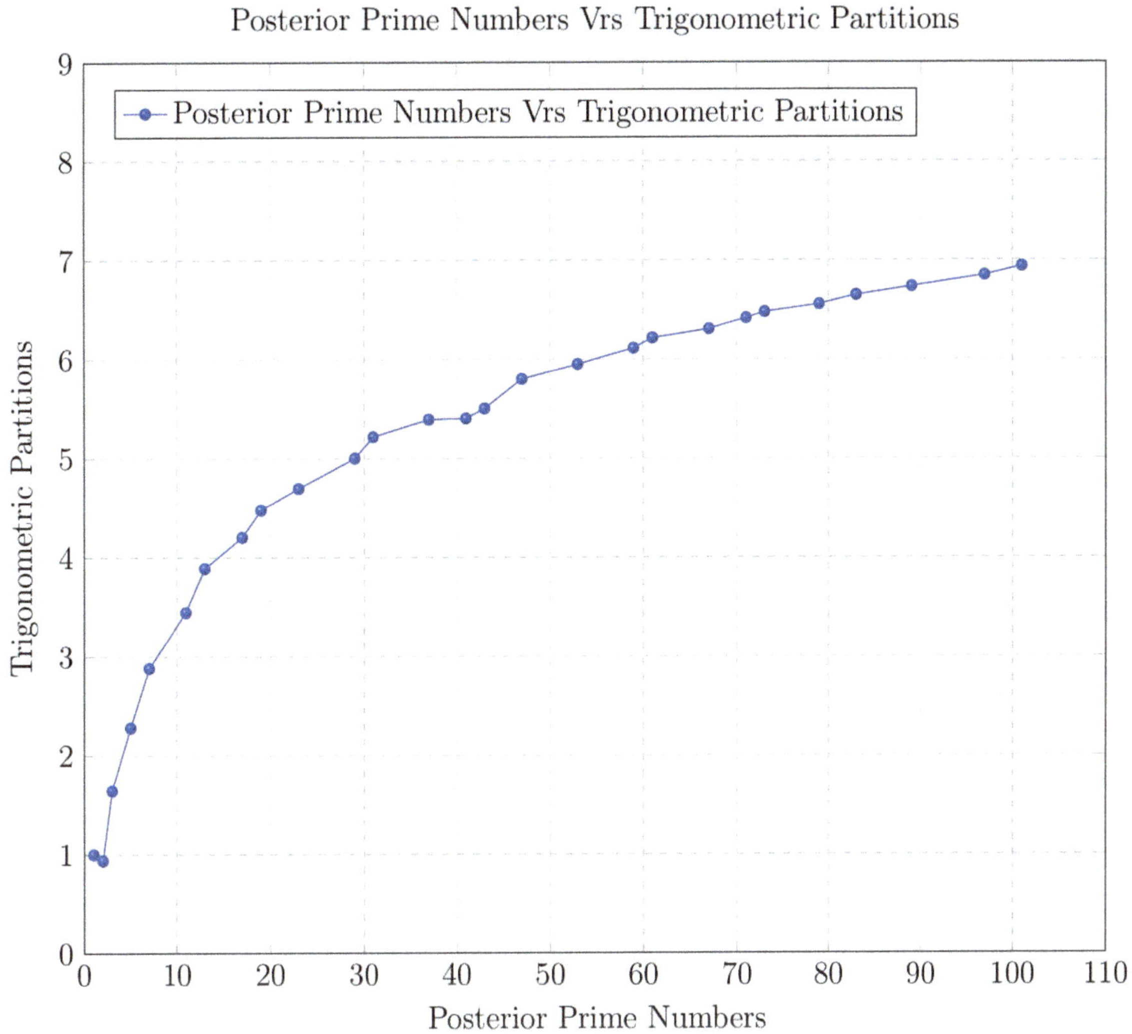

Figure 1.32: Posterior Prime Numbers Vrs Trigonometric Partitions, at 60 degrees, s=1,f=1.

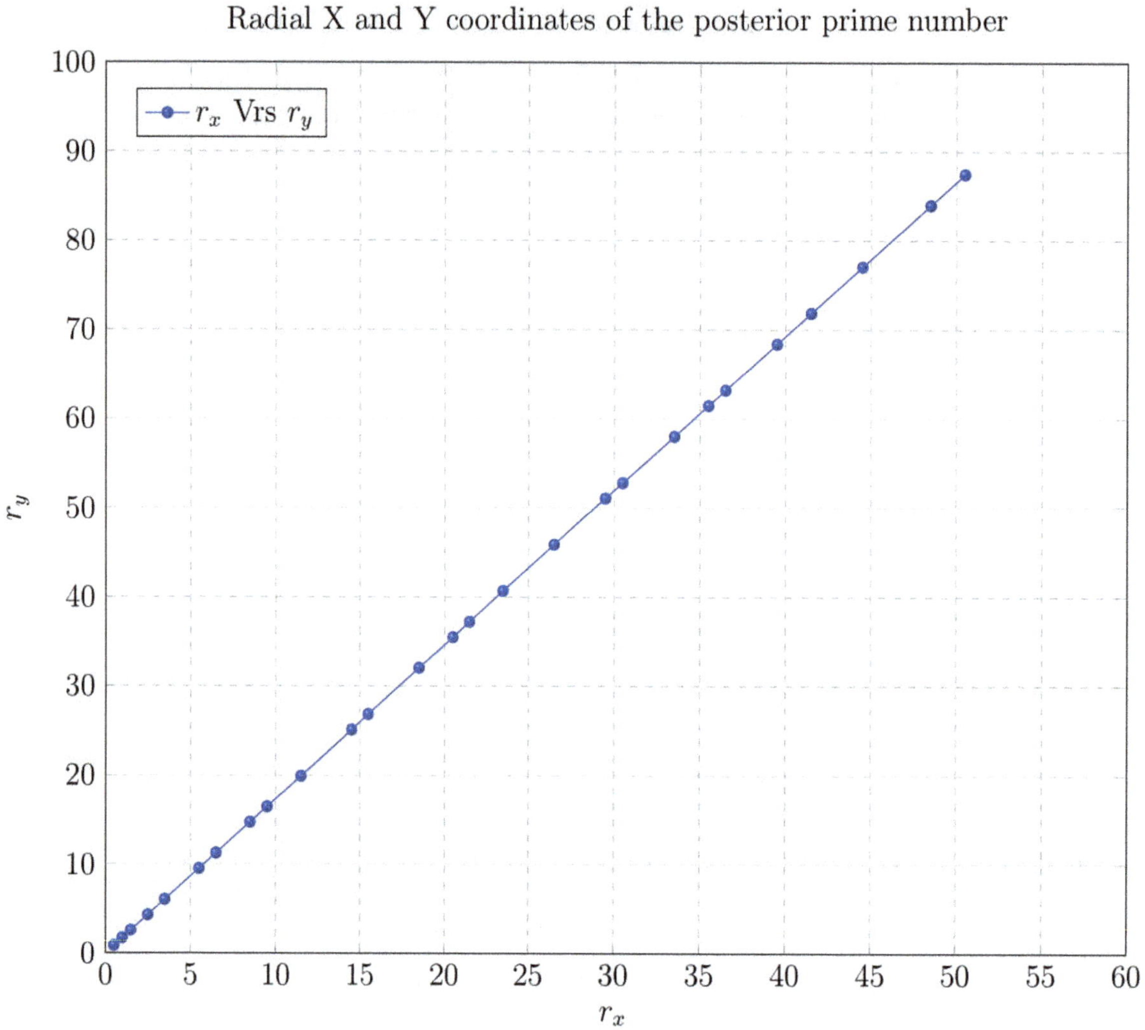

Figure 1.33: Radial X and Y coordinates of the posterior prime number, at 60 degrees, s=1,f=1.

Table 1.11: Prime Numbers Formula for +60 Degrees.

Factc	f=	0.86603	m=	2	Q=	1	k=	1	s=	2
Qty	**Posterior prime number parti- tions** p_1	**Previous Prime Num- ber** r_1	**Module** ΔL	**Module** $\sum\Delta L$	**Posterior prime number** r_3	**Spiral Angle** θ **Rads**	**Spiral An- gle** θ **Degs**	**Calculated Posterior Prime Number** r_3	**Coord X Real Posterior Prime Number** r_{x3}	**Coord Y Real Posterior Prime Number** r_{y3}
1	1	1	0	0	1.00	1.047198	60.00	1.000000	0.500000	0.866025
2	1.6138200	1	3	3	4.00	1.047198	60.00	4.000000	2.000000	3.464101
3	3.1076300	4	5	8	9.00	1.047198	60.00	9.000000	4.500000	7.794228
4	4.4232600	9	16	24	25.00	1.047198	60.00	25.000000	12.500000	21.650635
5	5.6443000	25	24	48	49.00	1.047198	60.00	49.000000	24.500000	42.435245
6	6.7817000	49	72	120	121.00	1.047198	60.00	121.000000	60.500000	104.789074
7	7.6750000	121	48	168	169.00	1.047198	60.00	169.000000	84.500000	146.358294
8	8.3030000	169	120	288	289.00	1.047198	60.00	289.000000	144.500000	250.281342
9	8.8500000	289	72	360	361.00	1.047198	60.00	361.000000	180.500000	312.635171
10	9.2860000	361	168	528	529.00	1.047198	60.00	529.000000	264.500000	458.127438
11	9.8960000	529	312	840	841.00	1.047198	60.00	841.000000	420.500000	728.327364
12	10.3300000	841	120	960	961.00	1.047198	60.00	961.000000	480.500000	832.250413
13	10.68000000	961	408	1368	1369.00	1.047198	60.00	1369.000000	684.500000	1185.588778
14	11.08000000	1369	312	1680	1681.00	1.047198	60.00	1681.000000	840.500000	1455.788704
15	11.30000000	1681	168	1848	1849.00	1.047198	60.00	1849.000000	924.500000	1601.280972
16	11.50000000	1849	360	2208	2209.00	1.047198	60.00	2209.000000	1104.500000	1913.050117
17	11.80000000	2209	600	2808	2809.00	1.047198	60.00	2809.000000	1404.500000	2432.665359
18	12.13000000	2809	672	3480	3481.00	1.047198	60.00	3481.000000	1740.500000	3014.634431
19	12.30000000	3481	240	3720	3721.00	1.047198	60.00	3721.000000	1860.500000	3222.480527
20	12.50000000	3721	768	4488	4489.00	1.047198	60.00	4489.000000	2244.500000	3887.588037
21	12.70000000	4489	552	5040	5041.00	1.047198	60.00	5041.000000	2520.500000	4365.634060
22	12.90000000	5041	288	5328	5329.00	1.047198	60.00	5329.000000	2664.500000	4615.049377
23	13.00000000	5329	912	6240	6241.00	1.047198	60.00	6241.000000	3120.500000	5404.864545
24	13.20000000	6241	648	6888	6889.00	1.047198	60.00	6889.000000	3444.500000	5966.049007
25	13.40000000	6889	1032	7920	7921.00	1.047198	60.00	7921.000000	3960.500000	6859.787224
26	13.55000000	7921	1488	9408	9409.00	1.047198	60.00	9409.000000	4704.500000	8148.433024
27	13.90000000	9409	792	10200	10201.00	1.047198	60.00	10201.000000	5100.500000	8834.325144

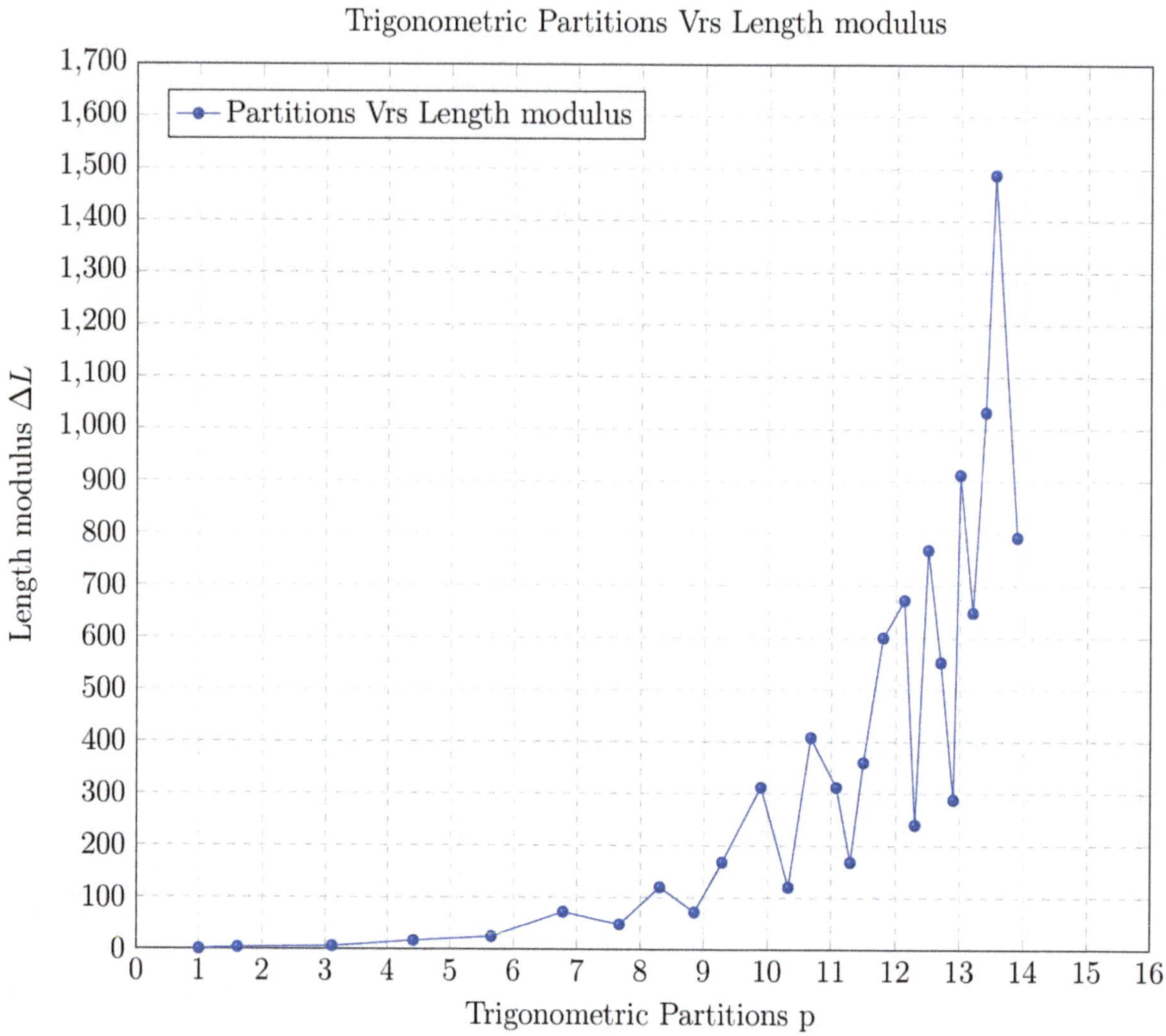

Figure 1.34: Trigonometric Partitions Vrs Length modulus, at 60 degrees, s=2,f=0.86603.

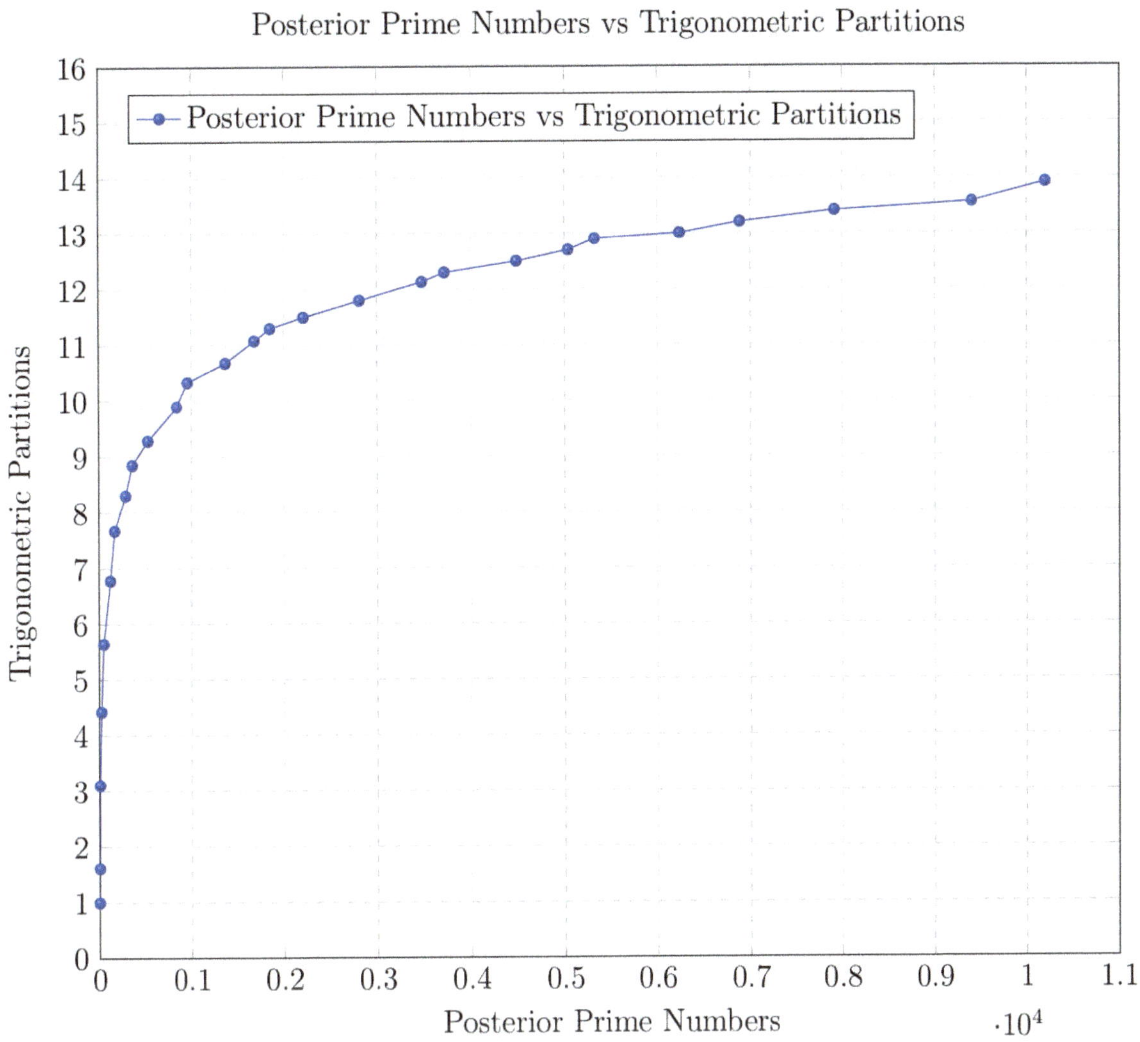

Figure 1.35: Posterior Prime Numbers Vrs Trigonometric Partitions, at 60 degrees, s=2, f=0.86603.

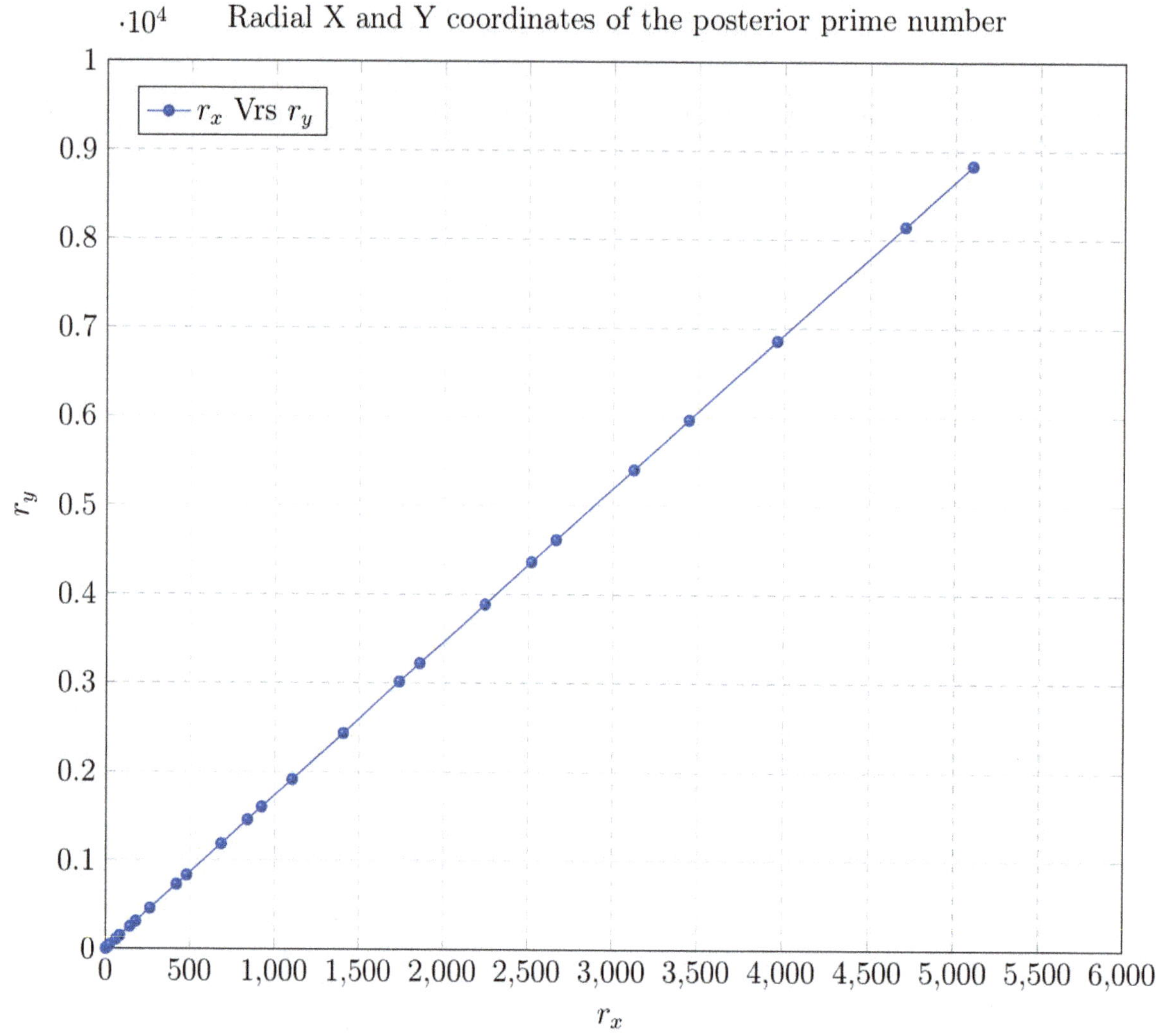

Figure 1.36: Radial X and Y coordinates of the posterior prime number, at 60 degrees, s=2, f=0.86603.

With the initial angle constant.;$\theta = 90^\circ = \pi/2$;The Trigonometric Partitions are calculated by trial and error;with f=0;$r_3(p) = \sqrt{[\Delta L * e^{[2^{(p-1)} * \sin(\pi * k/2^{(p)}) - \pi * k/2]} + \sqrt{(r_1^2 + f^2)}]^2 - f^2]}$ with f=0;$r_3(p) = \sqrt{[\Delta L * e^{[2^{(p-1)} \sin(\pi * k/2^{(p)}) - \pi * k/2]} + \sqrt{(r_1^2 + f^2)}]^2 - f^2]}$, finally, the equation will be: $r_3(p) = \Delta L * e^{[2^{(p-1)} * \sin[(\pi * k/2^{(p)}) - \pi\ k/2]} + r_1$.

Table 1.12: Prime Numbers Formula for $\theta_0 = +90$ Degrees.

Factc	f=	0	m=	1	Q=	1	k=	1	s=	1
Qty	**Posterior prime number partitions** p_1	**Previou Prime Number** r_1	**Module** ΔL	**Module** $\sum \Delta L$	**Posterior prime number** r_3	**Spiral Angle** θ **Rads**	**Spiral Angle** θ **Degs**	**Calculated Posterior Prime Number** r_3	**Coord X Real Posterior Prime Number** r_{x3}	**Coord Y Real Posterior Prime Number** r_{y3}
1	1	*1*	*0*	0	1	1.570796	90.000000	1.000000000000000000000	0.000000	1.000000
2	*25*	*1*	*1*	*1*	2	1.570796	90.000000	2.000000000000000000000	0.000000	2.000000
3	*26*	*2*	*1*	*2*	3	1.570796	90.000000	3.000000000000000000000	0.000000	3.000000
4	*25*	*3*	*2*	*4*	5	1.570796	90.000000	5.000000000000000000000	0.000000	5.000000
5	*30*	*5*	*2*	*6*	7	1.570796	90.000000	7.000000000000000000000	0.000000	7.000000
6	*32*	*7*	*4*	*10*	11	1.570796	90.000000	11.000000000000000000000	0.000000	11.000000
7	*37*	*11*	*2*	*12*	13	1.570796	90.000000	13.000000000000000000000	0.000000	13.000000
8	*40*	*13*	*4*	*16*	17	1.570796	90.000000	17.000000000000000000000	0.000000	17.000000
9	*43*	*17*	*2*	*18*	19	1.570796	90.000000	19.000000000000000000000	0.000000	19.000000
10	*48*	*19*	*4*	*22*	23	1.570796	90.000000	23.000000000000000000000	0.000000	23.000000
11	*49*	*23*	*6*	*28*	29	1.570796	90.000000	29.000000000000000000000	0.000000	29.000000
12	*52*	*29*	*2*	*30*	31	1.570796	90.000000	31.000000000000000000000	0.000000	31.000000
13	*56*	*31*	*6*	*36*	37	1.570796	90.000000	37.000000000000000000000	0.000000	37.000000
14	*59*	*37*	*4*	*40*	41	1.570796	90.000000	41.000000000000000000000	0.000000	41.000000
15	*60*	*41*	*2*	*42*	43	1.570796	90.000000	43.000000000000000000000	0.000000	43.000000
16	*65*	*43*	*4*	*46*	47	1.570796	90.000000	47.000000000000000000000	0.000000	47.000000
17	*67*	*47*	*6*	*52*	53	1.570796	90.000000	53.000000000000000000000	0.000000	53.000000
18	*69*	*53*	*6*	*58*	59	1.570796	90.000000	59.000000000000000000000	0.000000	59.000000
19	*72*	*59*	*2*	*60*	61	1.570796	90.000000	61.000000000000000000000	0.000000	61.000000
20	*75*	*61*	*6*	*66*	67	1.570796	90.000000	67.000000000000000000000	0.000000	67.000000
21	*77*	*67*	*4*	*70*	71	1.570796	90.000000	71.000000000000000000000	0.000000	71.000000
22	*79*	*71*	*2*	*72*	73	1.570796	90.000000	73.000000000000000000000	0.000000	73.000000
23	*82*	*73*	*6*	*78*	79	1.570796	90.000000	79.000000000000000000000	0.000000	79.000000
24	*84*	*79*	*4*	*82*	83	1.570796	90.000000	83.000000000000000000000	0.000000	83.000000
25	*87*	*83*	*6*	*88*	89	1.570796	90.000000	89.000000000000000000000	0.000000	89.000000
26	*88*	*89*	*8*	*96*	97	1.570796	90.000000	97.000000000000000000000	0.000000	97.000000
27	*92*	*97*	*4*	*100*	101	1.570796	90.000000	101.000000000000000000000	0.000000	101.000000

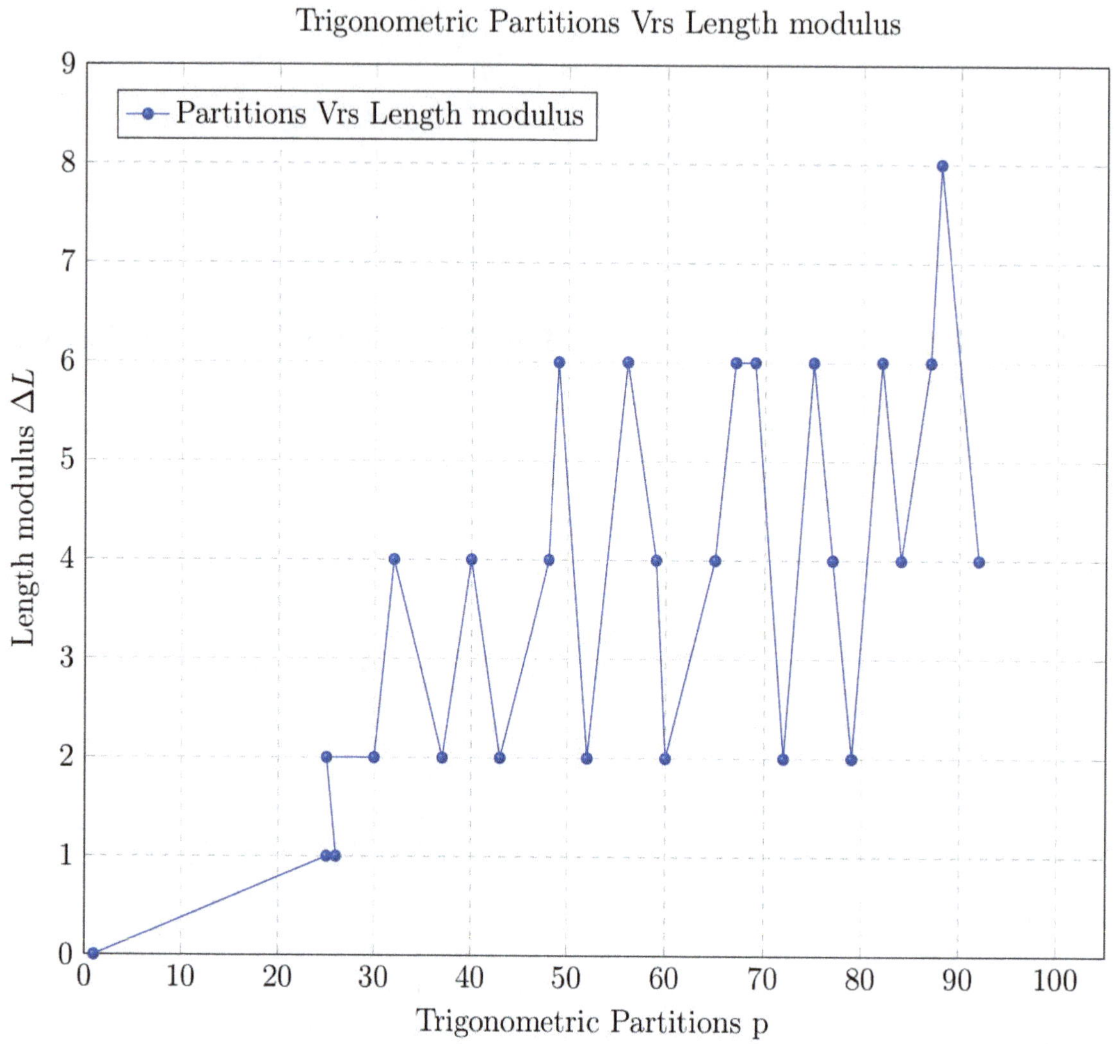

Figure 1.37: Trigonometric Partitions Vrs Length modulus, at 90 degrees, s=1,f=0.

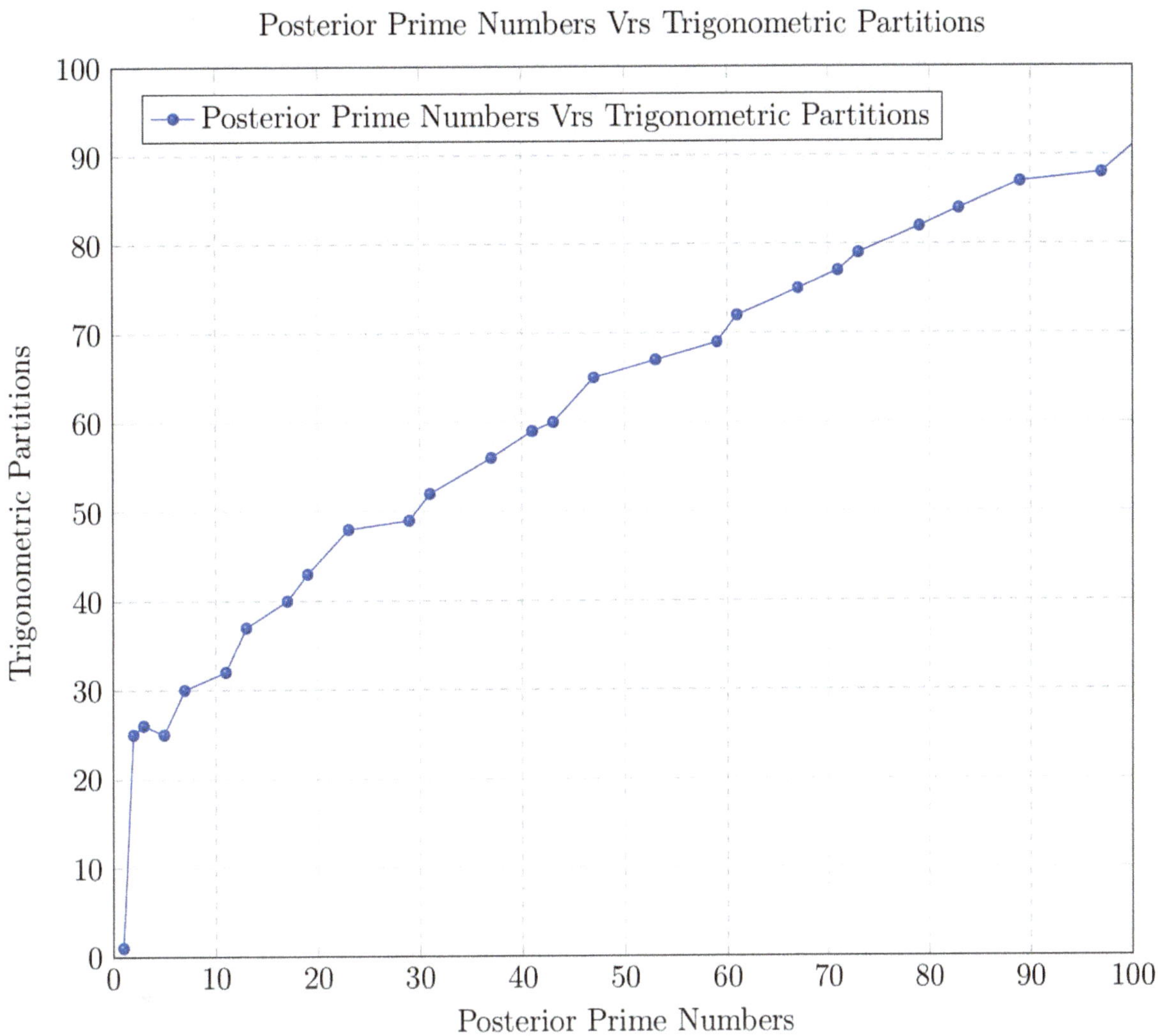

Figure 1.38: Posterior Prime Numbers Vrs Trigonometric Partitions, at 90 degrees, s=1,f=0.

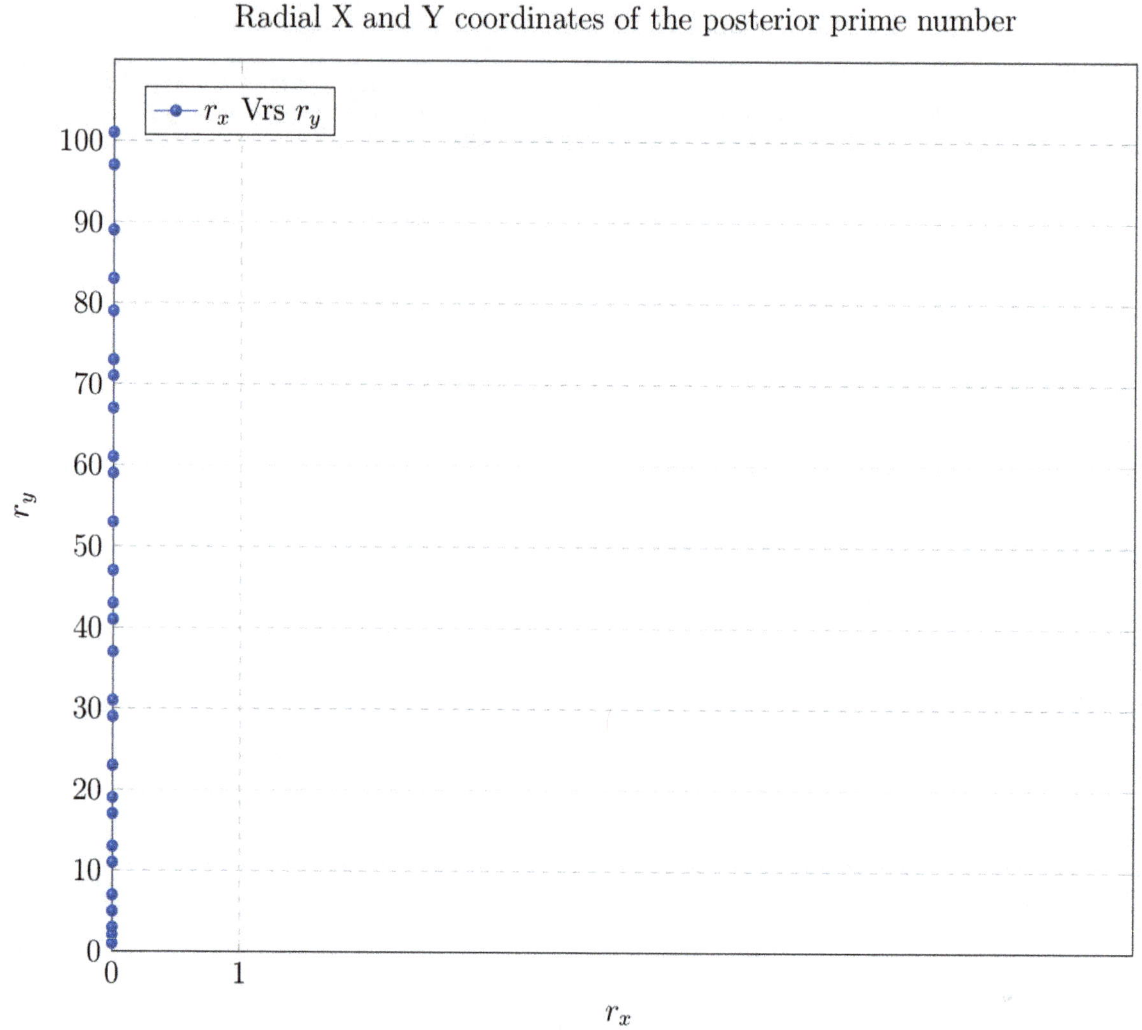

Figure 1.39: Radial X and Y coordinates of the posterior prime number, at 90 degrees, s=1,f=0.

1.3 Equations Of Prime Numbers In The Plane Of Real Numbers.

If we use these same equations in GeoGebra with the initial angle constant;

$\theta = 45° = \pi/4 = \pi/2^2$.In radians; with the trial and error calculations of the Trigonometric Partitions;

$r_3(p) = \sqrt{[\Delta L * e^{[2^{(p-1)}*\sin(\pi*k/2^{(p+1)})-\pi*k/4]} + \sqrt{(r_1^2+f^2)}]^2 - f^2}$.

, The exponent s=1 is constant. Therefore, if we obtain the polar coordinates of this equation, both positive: **a**, and negative: **e**, we will obtain invisible dots, so that they are marked, they must be selected to show the trace in the 2D plane. Meanwhile, in the 3D plane, these dots of the two polar coordinates generate arcs or circles depending on their rotation or initial values of the selected spiral angles depending on the multiple **k** rotations. Remember, r_3 is the calculation of the posterior prime number, r_1 is the previous prime number, ΔL the growth of the radius with respect to the posterior prime number minus the previous one, **p** is the partitions of the posterior prime number, **f** is the height on the **Y** axis, **k** is the number of turns.

2D 3D

Figure 1.40: Graphics In The Plane Of Real Numbers.

In this graph, the **k** slider has been activated so that the polar coordinates rotate and generate a graph in the 2D and 3D plane.In this section r_3 is function of p; $r_3(p)$.If we use these same equations in GeoGebra with the constant initial angle;$\theta = 45^\circ = \pi/4$, in radians;With the trial and error calculations of the Trigonometric Partitions;

$r_3(p) = \sqrt{[\Delta L * e^{[2^{(p-1)} * \sin(\pi * k/2^{(p+1)}) - \pi * k/4]} + \sqrt{(r_1^2 + f^2)}]^2 - f^2]}.$

The exponent s=1 is constant, the polar coordinates a and e respectively are:

$r_{3x}(p) = \cos(\pi * k/4) * \sqrt{[\Delta L * e^{[2^{(p-1)} * \sin(\pi * k/2^{(p+1)}) - \pi * k/4]} + \sqrt{(r_1^2 + f^2)}]^2 - f^2]}.$

$r_{3y}(p) = \sin(\pi * k/4) * \sqrt{[\Delta L * e^{[2^{(p-1)} * \sin(\pi * k/2^{(p+1)}) - \pi * k/4]} + \sqrt{(r_1^2 + f^2)}]^2 - f^2]}.$

If we use these same equations in GeoGebra with the constant initial angle;

$\theta = 45^\circ = \pi/4$; With the trial and error calculations of the Trigonometric Partitions;

$r_3(p) = \sqrt{[\Delta L * e^{[2^{(p-1)} * \sin(\pi * k/2^{(p+1)}) - \pi * k/4]} + \sqrt{(r_1^2 + f^2)}]^2 - f^2]}$,green color graph

$r_3(p) = -r_3(p) = -\sqrt{[\Delta L * e^{[2^{(p-1)} * \sin(\pi * k/2^{(p+1)}) - \pi * k/4]} + \sqrt{(r_1^2 + f^2)}]^2 - f^2]}$, orange graph,

The exponent of **n** with real numbers; s=1 constant.

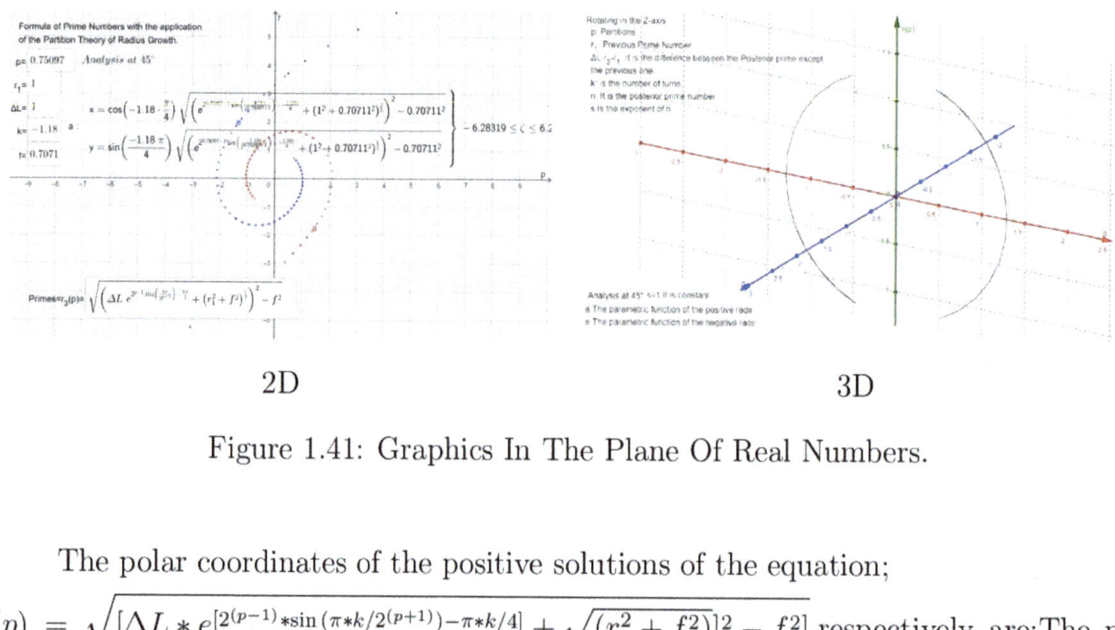

2D 3D

Figure 1.41: Graphics In The Plane Of Real Numbers.

The polar coordinates of the positive solutions of the equation;

$r_3(p) = \sqrt{[\Delta L * e^{[2^{(p-1)} * \sin(\pi * k/2^{(p+1)}) - \pi * k/4]} + \sqrt{(r_1^2 + f^2)}]^2 - f^2]}$,respectively are:The polar coordinates of the negative solutions of the equation;

$r_{3x}(p) = \cos(\pi * k/4) * \sqrt{[\Delta L * e^{[2^{(p-1)} * \sin(\pi * k/2^{(p+1)}) - \pi * k/4]} + \sqrt{(r_1^2 + f^2)}]^2 - f^2]}.$

$r_{3y}(p) = \sin(\pi * k/4) * \sqrt{[\Delta L * e^{[2^{(p-1)} * \sin(\pi * k/2^{(p+1)}) - \pi * k/4]} + \sqrt{(r_1^2 + f^2)}]^2 - f^2]}.$

$r_3(p) = -\sqrt{[\Delta L * e^{[2^{(p-1)} * \sin(\pi * k/2^{(p+1)}) - \pi * k/4]} + \sqrt{(r_1^2 + f^2)}]^2 - f^2]},$

respectively are:

$$r_{3x}(p) = -\cos(\pi * k/4) * \sqrt{[\Delta L * e^{[2^{(p-1)} * \sin(\pi * k/2^{(p+1)}) - \pi * k/4]} + \sqrt{(r_1^2 + f^2)}]^2 - f^2}.$$

$$r_{3y}(p) = -\sin(\pi * k/4) * \sqrt{[\Delta L * e^{[2^{(p-1)} * \sin(\pi * k/2^{(p+1)}) - \pi * k/4]} + \sqrt{(r_1^2 + f^2)}]^2 - f^2}.$$

The graphics and coordinates are the same in the initial analysis at 45°, $r_{3x} = r_{3y}$, in each positive or negative equation of r_3.

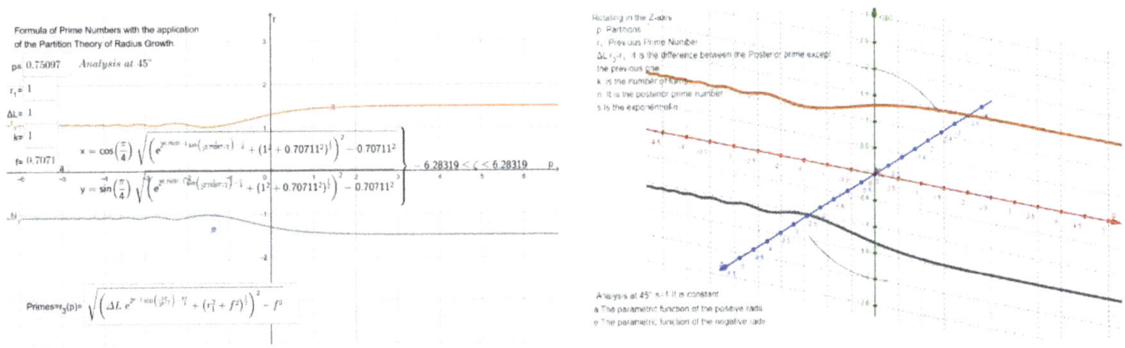

2D 3D

Figure 1.42: Graphics In The Plane Of Real Numbers.

The following graphic is where all the features in the real plane have been selected in GeoGebra.

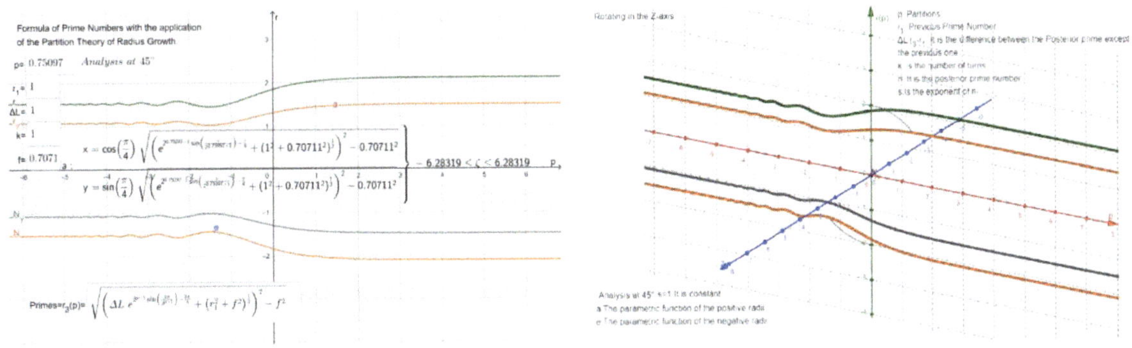

2D 3D

Figure 1.43: Graphics In The Plane Of Real Numbers.

If we use these same equations in GeoGebra with the constant initial angle; $\theta = 60^\circ = \pi/3$,in radians; With the trial and error calculations of the Trigonometric Partitions; $r_3(p) = \sqrt{[\Delta L * e^{[2^{(p-1)}*\sin(\pi*k/(3*2^{(p+1)}))-\pi*k/3]} + \sqrt{(r_1^2+f^2)}]^2 - f^2]}$,

The exponent s=1 is constant. Therefore, if we obtain the polar coordinates of this equation,

both positive: **a**, and negative: **e**, we will obtain invisible dots, so that they are marked, they must be selected to show the trace in the 2D plane. Meanwhile, in the 3D plane, these dots of the two polar coordinates generate arcs or circles depending on their rotation or initial values of the selected spiral angles depending on the multiple **k** rotations. Remembering, r_3 is the calculation of the posterior prime number,r_1 is the previous prime number,ΔL the growth of the radius concerning the posterior prime number minus the previous one, **p** is the partitions of the posterior prime number, **f** is the height in the **Y** axis, **k** the number of turns.

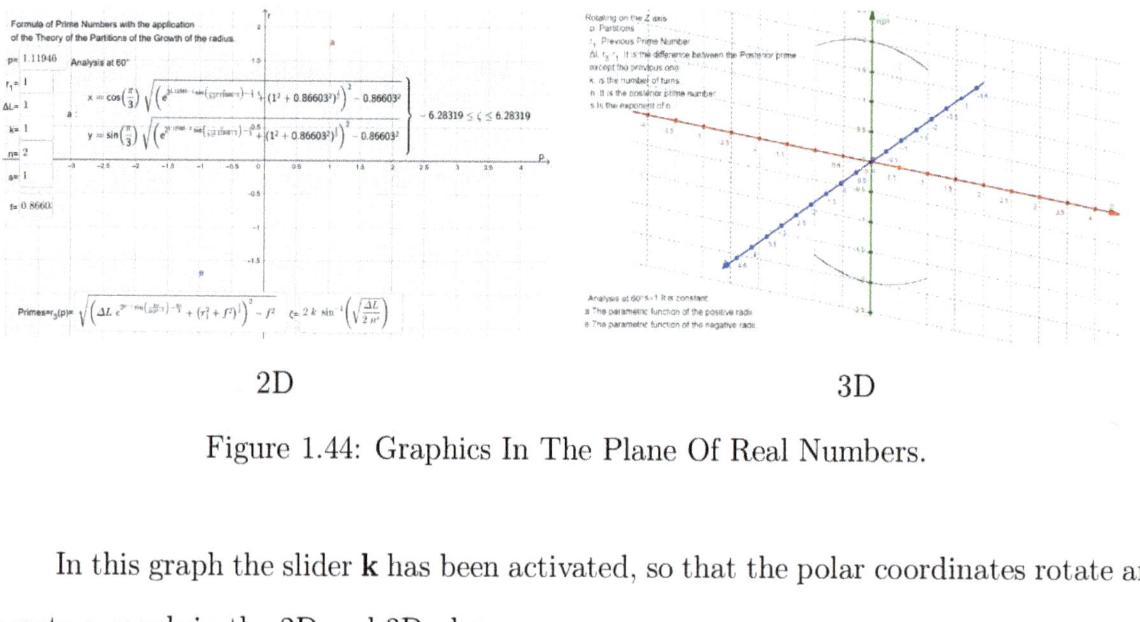

2D 3D

Figure 1.44: Graphics In The Plane Of Real Numbers.

In this graph the slider **k** has been activated, so that the polar coordinates rotate and generate a graph in the 2D and 3D plane.

If we use these same equations in GeoGebra with the constant initial angle; $\theta = 60^\circ = \pi/3$; With the trial and error calculations of the Trigonometric Partitions;

$r_3(p) = \sqrt{[\Delta L * e^{[2^{(p-1)}*\sin(\pi*k/(3*2^{(p+1)}))-\pi*k/3]} + \sqrt{(r_1^2+f^2)}]^2 - f^2]}$,

s=1 constant, the polar coordinates **a** and **e** respectively are:

$r_{3x}(p) = \cos(\pi*k/3) * \sqrt{[\Delta L * e^{[2^{(p-1)}*\sin(\pi*k/(3*2^{(p+1)}))-\pi*k/3]} + \sqrt{(r_1^2+f^2)}]^2 - f^2]}$.

$$r_{3y}(p) = \sin(\pi * k/3) * \sqrt{[\Delta L * e^{[2^{(p-1)} * \sin(\pi * k/(3 * 2^{(p+1)})) - \pi * k/3]} + \sqrt{(r_1^2 + f^2)}]^2 - f^2}.$$

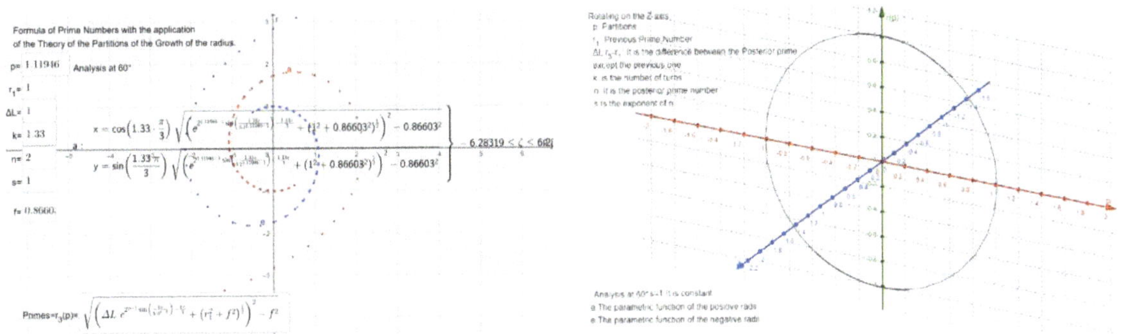

2D 3D

Figure 1.45: Graphics In The Plane Of Real Numbers.

If we use these same equations in GeoGebra with the constant initial angle;

$\theta = 60° = \pi/3$;With the trial and error calculations of the Trigonometric Partitions;

$r_3(p) = +r_3(p) = \sqrt{[\Delta L * e^{[2^{(p-1)} * \sin(\pi * k/(3 * 2^{(p+1)})) - \pi * k/3]} + \sqrt{(r_1^2 + f^2)}]^2 - f^2}$,

green color graph

$r_3(p) = -r_3(p) = -\sqrt{[\Delta L * e^{[2^{(p-1)} * \sin(\pi * k/(3 * 2^{(p+1)})) - \pi * k/3]} + \sqrt{(r_1^2 + f^2)}]^2 - f^2}$,orange color graph, The exponent of **n** with real numbers; s=1 constant.

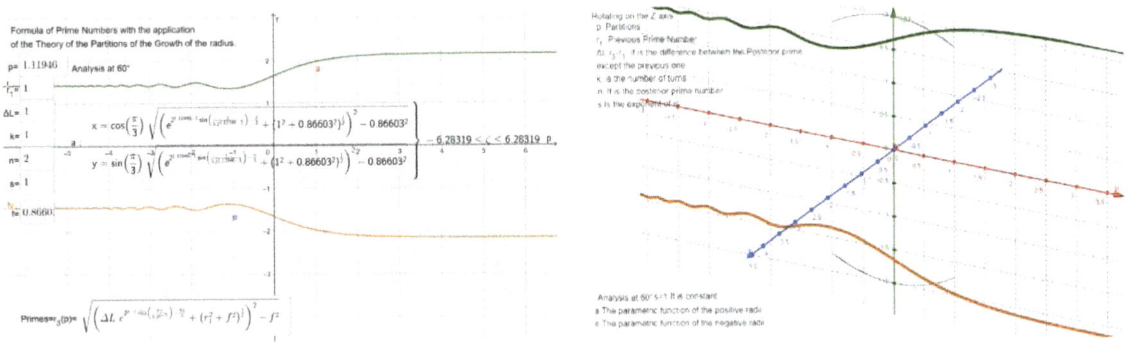

2D 3D

Figure 1.46: Graphics In The Plane Of Real Numbers.

The polar coordinates of the positive solutions of the equation;

$r_3(p) = \sqrt{[\Delta L * e^{[2^{(p-1)} * \sin(\pi * k/(3 * 2^{(p+1)})) - \pi * k/3]} + \sqrt{(r_1^2 + f^2)}]^2 - f^2}$,respectively are:

$$r_{3x}(p) = \cos(\pi * k/3) * \sqrt{[\Delta L * e^{[2^{(p-1)} * \sin(\pi * k/(3 * 2^{(p+1)})) - \pi * k/3]} + \sqrt{(r_1^2 + f^2)}]^2 - f^2}.$$

$$r_{3y}(p) = \sin(\pi * k/3) * \sqrt{[\Delta L * e^{[2^{(p-1)} * \sin(\pi * k/(3 * 2^{(p+1)})) - \pi * k/3]} + \sqrt{(r_1^2 + f^2)}]^2 - f^2}.$$

The polar coordinates of the negative solutions of the equation;

$$r_3(p) = -\sqrt{[\Delta L * e^{[2^{(p-1)} * \sin(\pi * k/(3*2^{(p+1)})) - \pi * k/3]} + \sqrt{(r_1^2 + f^2)}]^2 - f^2}.$$

respectively are:

$$r_{3x}(p) = -\cos(\pi * k/3) * \sqrt{[\Delta L * e^{[2^{(p-1)} * \sin(\pi * k/(3*2^{(p+1)})) - \pi * k/3]} + \sqrt{(r_1^2 + f^2)}]^2 - f^2}.$$

$$r_{3y}(p) = -\sin(\pi * k/3) * \sqrt{[\Delta L * e^{[2^{(p-1)} * \sin(\pi * k/(3*2^{(p+1)})) - \pi * k/3]} + \sqrt{(r_1^2 + f^2)}]^2 - f^2}.$$

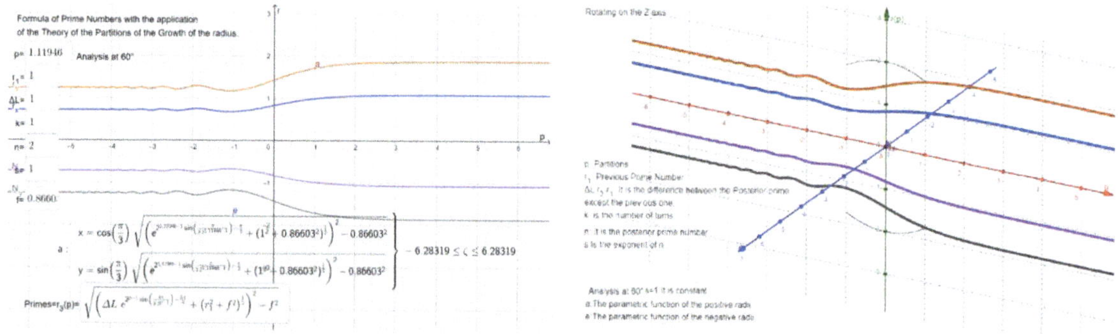

2D 3D

Figure 1.47: Graphics In The Plane Of Real Numbers.

The following graphic is where all the features in the real plane have been selected in GeoGebra.

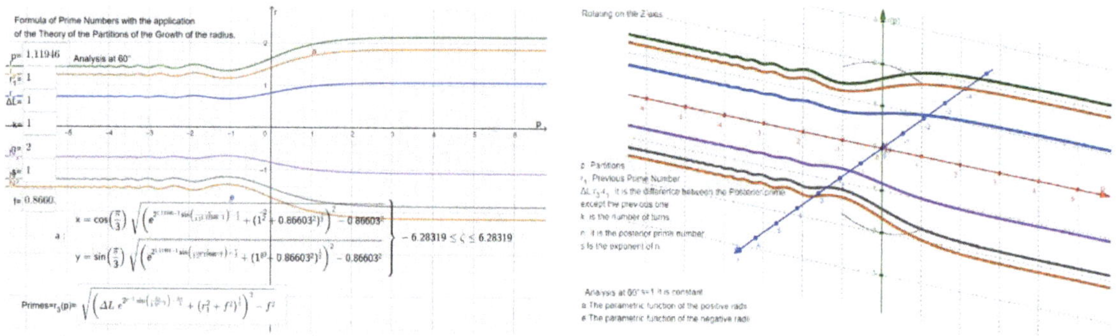

2D 3D

Figure 1.48: Graphics In The Plane Of Real Numbers.

If we use these same equations in GeoGebra with its equation for the variable initial spiral angle in radians; θ, the height f, the prime before r_1,the growth of the radius ΔL, the real posterior prime number n, and the exponents; all variables, also with the variable values of the Trigonometric Partitions. Then:

$r_3(p) = \sqrt{[\Delta L * e^{2(p-1)*\sin(\theta/2^{(p-1)})-\theta} + \sqrt{(r_1^2+f^2)}]^2 - f^2}$,Therefore, if we obtain the polar coordinates of this equation, both positive: **a**, and negative: **e**, we will obtain invisible dots, so that they are marked, they must be selected to show the trace in the 2D plane. Meanwhile, in the 3D plane these dots of the two polar coordinates generate rotating arcs or circles depending on their rotation or initial values of the selected initial spiral angles depending on the multiple of **k** rotations, the height **f** on the **Y** axis, the numbers real number **n** and the exponent of the real numbers **s** in the Cartesian plane of the real numbers. The spiral angle ζ in radians.

$\zeta = k * \arctan(f/n^s) = \theta = k * \arcsin[f/\sqrt{(f^2+n^{2s})}]$.

In this graph, all the sliders have been activated, so that the polar coordinates **a**, and **e** rotate and generate a graph in the 2D and 3D planes. If we use these same equations in GeoGebra with all the variables and their active sliders in GeoGebra, with the variable values of the Trigonometric Partitions;

$r_3(p) = \sqrt{[\Delta L * e^{2(p-1)*\sin(\theta/2^{(p-1)})-\theta} + \sqrt{(r_1^2+f^2)}]^2 - f^2}$,the polar coordinates **a** and **e** respectively are:

$r_{3x}(p) = \cos(\theta) * \sqrt{[\Delta L * e^{2(p-1)*\sin(\theta/2^{(p-1)})-\theta} + \sqrt{(r_1^2+f^2)}]^2 - f^2}$.

$r_{3y}(p) = \sin(\theta) * \sqrt{[\Delta L * e^{2(p-1)*\sin(\theta/2^{(p-1)})-\theta} + \sqrt{(r_1^2+f^2)}]^2 - f^2}$.

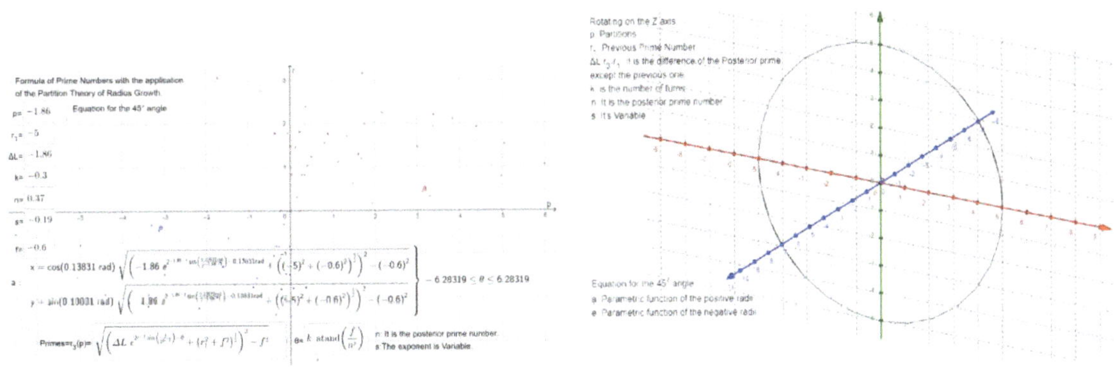

2D 3D

Figure 1.49: Graphics In The Plane Of Real Numbers.

If we use these same equations in GeoGebra after a certain time, with **k** and **f** variables and their active sliders in GeoGebra, with the trial and error calculations of the Trigonometric

Partitions;

$$r_3(p) = \sqrt{[\Delta L * e^{2^{(p-1)} * \sin(\theta/2^{(p-1)}) - \theta} + \sqrt{(r_1^2 + f^2)}]^2 - f^2},$$

the polar coordinates **a** and **e** respectively are:

$$r_{3x}(p) = \cos(\theta) * \sqrt{[\Delta L * e^{2^{(p-1)} * \sin(\theta/2^{(p-1)}) - \theta} + \sqrt{(r_1^2 + f^2)}]^2 - f^2}.$$

$$r_{3y}(p) = \sin(\theta) * \sqrt{[\Delta L * e^{2^{(p-1)} * \sin(\theta/2^{(p-1)}) - \theta} + \sqrt{(r_1^2 + f^2)}]^2 - f^2}.$$

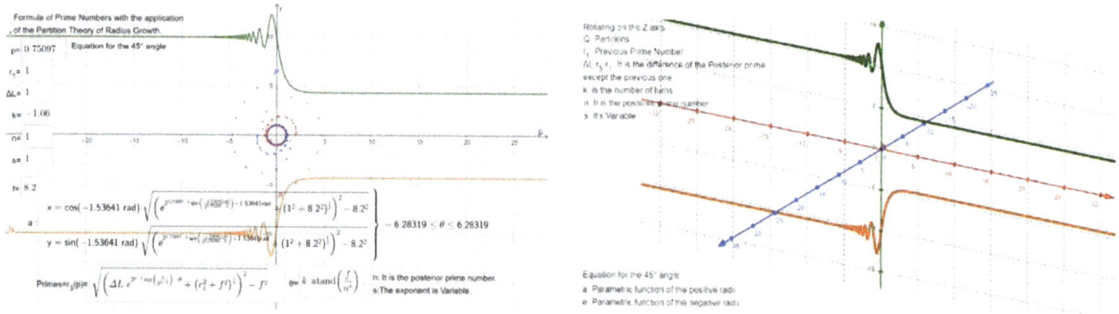

2D 3D

Figure 1.50: Graphics In The Plane Of Real Numbers.

If we use these same equations in GeoGebra with all the active sliders varying in time, with the trial and error calculations of the Trigonometric Partitions;

$r_3(p) = +r_3(p) = \sqrt{[\Delta L * e^{2^{(p-1)} * \sin(\theta/2^{(p-1)}) - \theta} + \sqrt{(r_1^2 + f^2)}]^2 - f^2}$, green color graph.

$r_3(p) = -r_3(p) = -\sqrt{[\Delta L * e^{2^{(p-1)} * \sin(\theta/2^{(p-1)}) - \theta} + \sqrt{(r_1^2 + f^2)}]^2 - f^2}$,orange color graph.

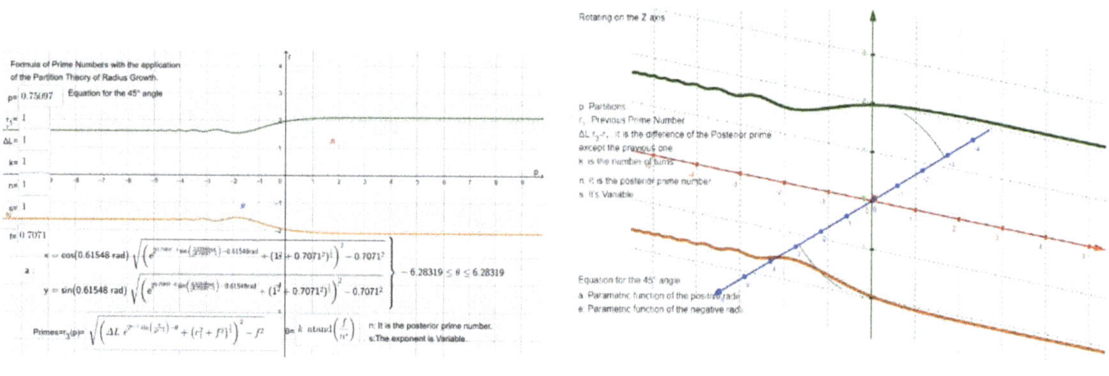

2D 3D

Figure 1.51: Graphics In The Plane Of Real Numbers.

The polar coordinates of the positive solutions of the prime numbers equation;

$r_3(p) = \sqrt{[\Delta L * e^{2^{(p-1)} * \sin(\theta/2^{(p-1)}) - \theta} + \sqrt{(r_1^2 + f^2)}]^2 - f^2}$ respectively are:

$$r_{3x}(p) = \cos(\theta) * \sqrt{[\Delta L * e^{2^{(p-1)} * \sin(\theta/2^{(p-1)}) - \theta} + \sqrt{(r_1^2 + f^2)}]^2 - f^2}.$$

$$r_{3y}(p) = \sin(\theta) * \sqrt{[\Delta L * e^{2^{(p-1)} * \sin(\theta/2^{(p-1)}) - \theta} + \sqrt{(r_1^2 + f^2)}]^2 - f^2}.$$

The polar coordinates of the negative solutions of the prime numbers equation;

$r_3(p) = -\sqrt{[\Delta L * e^{2^{(p-1)} * \sin(\theta/2^{(p-1)}) - \theta} + \sqrt{(r_1^2 + f^2)}]^2 - f^2}$ respectively are:

$$r_{3x}(p) = -\cos(\theta) * \sqrt{[\Delta L * e^{2^{(p-1)} * \sin(\theta/2^{(p-1)}) - \theta} + \sqrt{(r_1^2 + f^2)}]^2 - f^2}.$$

$$r_{3y}(p) = -\sin(\theta) * \sqrt{[\Delta L * e^{2^{(p-1)} * \sin(\theta/2^{(p-1)}) - \theta} + \sqrt{(r_1^2 + f^2)}]^2 - f^2}.$$

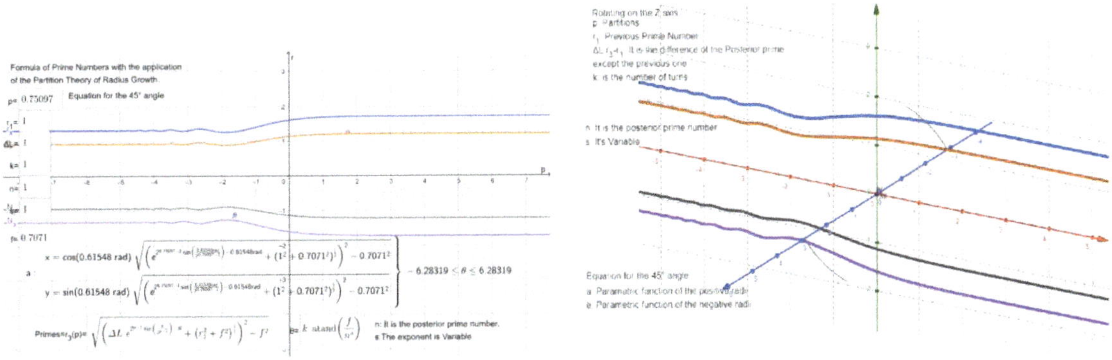

2D 3D

Figure 1.52: Graphics In The Plane Of Real Numbers.

The following graphic is where all the features in the real plane have been selected in GeoGebra.

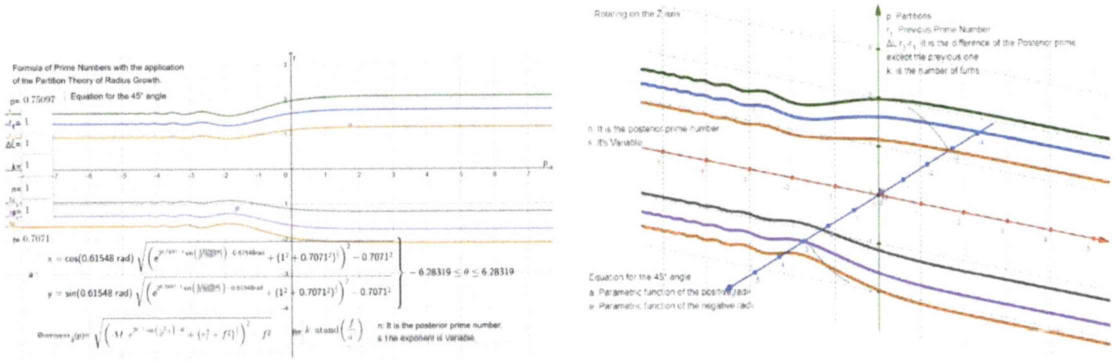

2D 3D

Figure 1.53: Graphics In The Plane Of Real Numbers.

If we use these same equations in GeoGebra with its equation for the variable initial spiral angle ζ, the height **f**, the previous prime number r_1, the growth of the radius ΔL, the real number **n**, and the exponent **s**; all variables, also with the variable values of the Trigonometric Partitions. Then: $r_3(p) = \sqrt{[\Delta L * e^{2(p-1)*\sin(\zeta/2(p-1))-\zeta} + \sqrt{(r_1^2+f^2)}]^2 - f^2}$,Therefore, if we obtain the polar coordinates of this equation, both positive: **a**, and negative: **e**, we will obtain invisible dots, so that they are marked, they must be selected to show the trace in the 2D plane. Meanwhile, in the 3D plane, these dots of the two polar coordinates generate rotating arcs or circles depending on their rotation or initial values of the selected initial spiral angles. Where,$\Delta r = 2 * r * \sin^2(\zeta/2)$.With these equations we can study the equation of Prime Numbers from two different points of view; therefore, the spiral angles will be:

$\zeta = 2 * k * \arcsin\sqrt{[\Delta L/(2 * n^s)]}$, or

$\theta = k * \arcsin\,[f/\sqrt{(f^2+n^{2s})}]$.Where **n** is the posterior prime, **k** is the turns or twists, **f** is the height in the **Y** axis, and $\Delta L = \Delta r$: the difference of the posterior prime minus the previous one. **S** is the exponent of the real numbers in the Cartesian number plane real. This is why I say that care must be taken with the equation to use for the angle of the prime number equation, such as Riemann's Z Function.

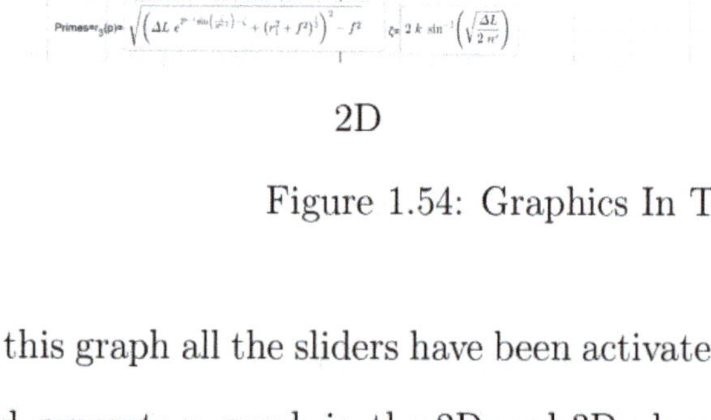

2D

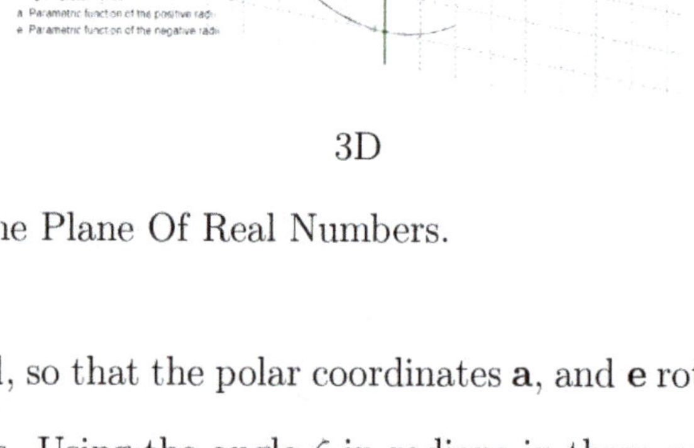

3D

Figure 1.54: Graphics In The Plane Of Real Numbers.

In this graph all the sliders have been activated, so that the polar coordinates **a**, and **e** rotate and generate a graph in the 2D and 3D planes. Using the angle ζ in radians in these equations. If we use these same equations in GeoGebra with all the variables and their sliders active in GeoGebra, with the variable values of the Trigonometric Partitions;

$r_3(p) = \sqrt{[\Delta L * e^{2(p-1)*\sin(\zeta/2^{(p-1)})-\zeta} + \sqrt{(r_1^2 + f^2)}]^2 - f^2}$,the polar coordinates **a** and **e** respectively are:

$r_{3x}(p) = \cos(\zeta) * \sqrt{[\Delta L * e^{2(p-1)*\sin(\zeta/2^{(p-1)})-\zeta} + \sqrt{(r_1^2 + f^2)}]^2 - f^2}$.

$r_{3y}(p) = \sin(\zeta) * \sqrt{[\Delta L * e^{2(p-1)*\sin(\zeta/2^{(p-1)})-\zeta} + \sqrt{(r_1^2 + f^2)}]^2 - f^2}$.

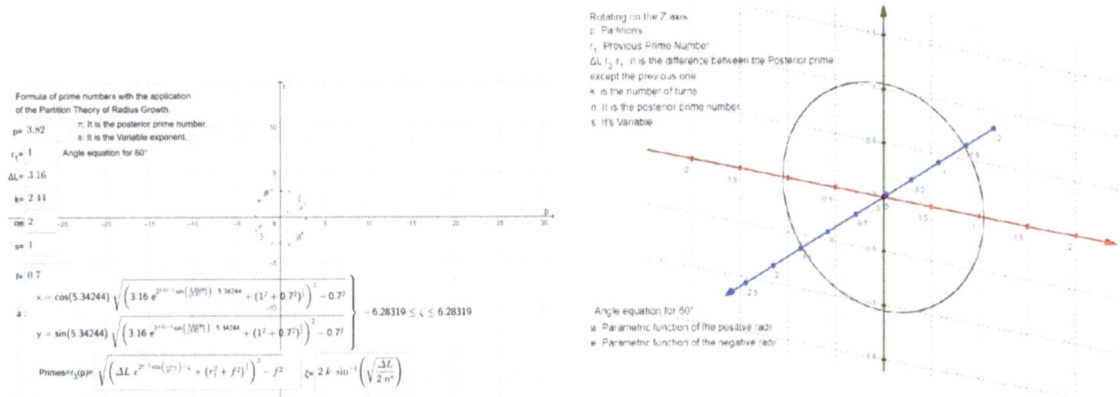

2D 3D

Figure 1.55: Graphics In The Plane Of Real Numbers.

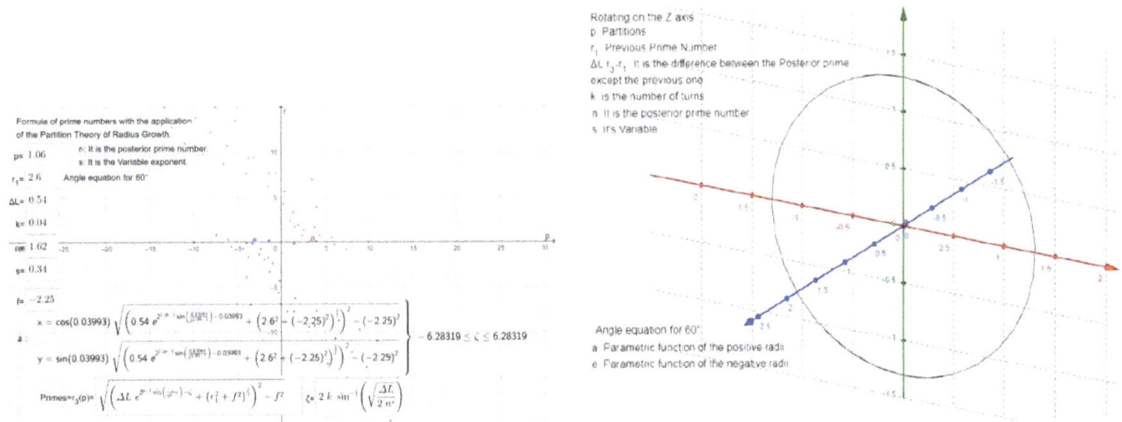

2D 3D

Figure 1.56: Graphics In The Plane Of Real Numbers.

If we use these same equations in GeoGebra after a certain time, with **k** and **f** variables and their active sliders in GeoGebra, with the trial and error calculations of the Trigonometric Partitions; $r_3(p) = \sqrt{[\Delta L * e^{2(p-1)*\sin(\zeta/2^{(p-1)})-\zeta} + \sqrt{(r_1^2 + f^2)}]^2 - f^2}$,the polar coordinates **a** and **e** respectively are:

$$r_{3x}(p) = \cos(\zeta) * \sqrt{[\Delta L * e^{2(p-1)*\sin(\zeta/2^{(p-1)})-\zeta} + \sqrt{(r_1^2 + f^2)}]^2 - f^2}.$$

$$r_{3y}(p) = \sin(\zeta) * \sqrt{[\Delta L * e^{2(p-1)*\sin(\zeta/2^{(p-1)})-\zeta} + \sqrt{(r_1^2 + f^2)}]^2 - f^2}.$$

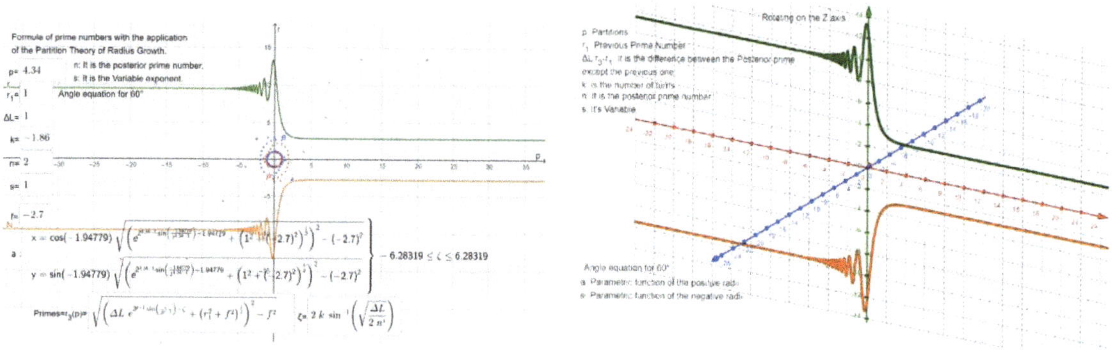

2D 3D

Figure 1.57: Graphics In The Plane Of Real Numbers.

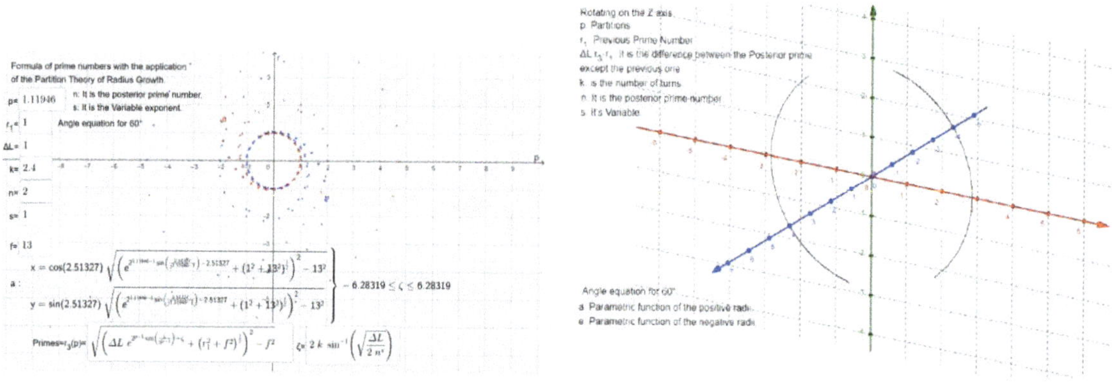

2D 3D

Figure 1.58: Graphics In The Plane Of Real Numbers.

If we use these same equations in GeoGebra with all the active sliders varying in time, with the trial and error calculations of the Trigonometric Partitions;

$r_{3x}(p) = +r_{3x}(p) = \cos(\zeta) * \sqrt{[\Delta L * e^{2(p-1)*\sin(\zeta/2^{(p-1)})-\zeta} + \sqrt{(r_1^2 + f^2)}]^2 - f^2}$,green color graph.

$r_{3x}(p) = -r_{3x}(p) = -\cos(\zeta) * \sqrt{[\Delta L * e^{2(p-1)*\sin(\zeta/2^{(p-1)})-\zeta} + \sqrt{(r_1^2 + f^2)}]^2 - f^2}$,Orange color graph.

The polar coordinates of the positive solutions of the prime number equation;

$r_3(p) = \sqrt{[\Delta L * e^{2(p-1)*\sin(\zeta/2^{(p-1)})-\zeta} + \sqrt{(r_1^2 + f^2)}]^2 - f^2}$,respectively are:

$r_{3x}(p) = \cos(\zeta) * \sqrt{[\Delta L * e^{2(p-1)*\sin(\zeta/2^{(p-1)})-\zeta} + \sqrt{(r_1^2 + f^2)}]^2 - f^2}$.

$r_{3y}(p) = -\sin(\zeta) * \sqrt{[\Delta L * e^{2(p-1)*\sin(\zeta/2^{(p-1)})-\zeta} + \sqrt{(r_1^2 + f^2)}]^2 - f^2}$.

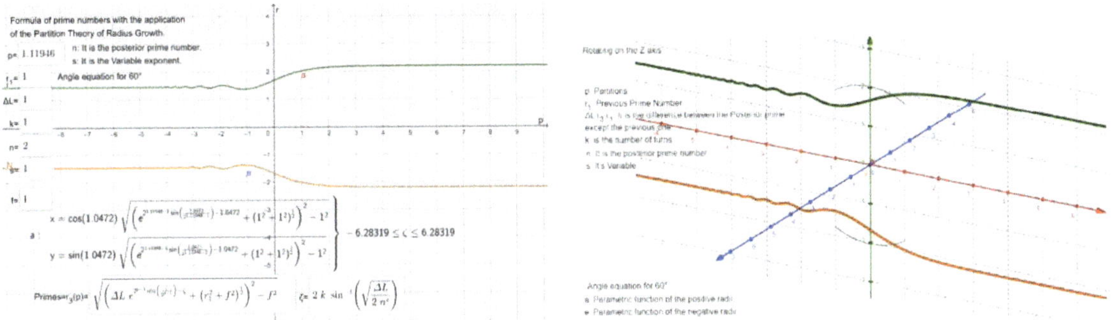

2D 3D

Figure 1.59: Graphics In The Plane Of Real Numbers.

The polar coordinates of the negative solutions of the prime number equation;

$r_3(p) = -\sqrt{[\Delta L * e^{2(p-1)*\sin(\zeta/2^{(p-1)})-\zeta} + \sqrt{(r_1^2 + f^2)}]^2 - f^2}$.respectively are:

$r_{3x}(p) = -\cos(\zeta) * \sqrt{[\Delta L * e^{2(p-1)*\sin(\zeta/2^{(p-1)})-\zeta} + \sqrt{(r_1^2 + f^2)}]^2 - f^2}$.

$r_{3y}(p) = -\sin(\zeta) * \sqrt{[\Delta L * e^{2(p-1)*\sin(\zeta/2^{(p-1)})-\zeta} + \sqrt{(r_1^2 + f^2)}]^2 - f^2}$

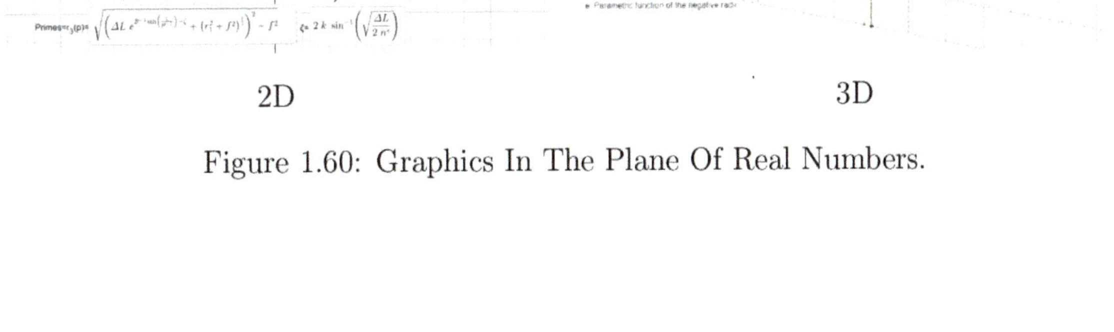

2D 3D

Figure 1.60: Graphics In The Plane Of Real Numbers.

The following graph is where all the features in the real plane have been selected in GeoGebra.

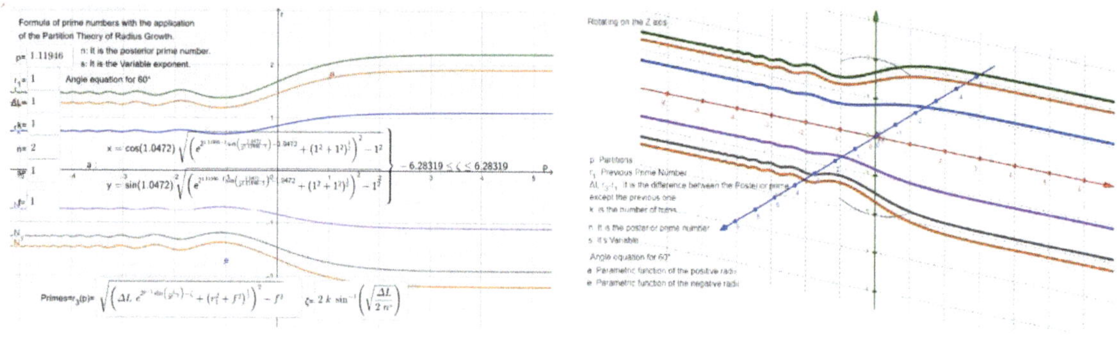

2D 3D

Figure 1.61: Graphics In The Plane Of Real Numbers

If we use these same formulations in GeoGebra with the constant initial angle; $\theta = 90° = \pi/2$, in radians; With the trial and error calculations of the Trigonometric Partitions;with f=0, $r_3(p) = \Delta L * e^{[2^{(p-1)}*\sin(\pi*k/2^p)-\pi*k/2]} + r_1$.

,The exponent s=1 constant, the polar coordinates **a** and **e** respectively are:

$r_{3x}(p) = \cos(\pi * k/2) * [\Delta L * e^{[2^{(p-1)}*\sin(\pi*k/2^p)-\pi*k/2]} + r_1]$.

$r_{3y}(p) = \sin(\pi * k/2) * [\Delta L * e^{[2^{(p-1)}*\sin(\pi*k/2^p)-\pi*k/2]} + r_1]$.

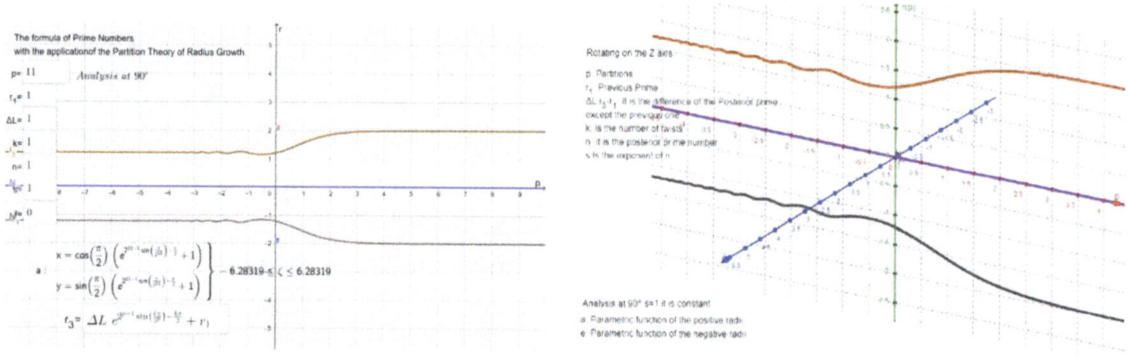

2D 3D

Figure 1.62: Graphics In The Plane Of Real Numbers.

1.4 Equations Of Prime Numbers In The Plane Of Complex Numbers.

If we use the same equations with an initial analysis of 45°; in the complex plane with GeoGebra and its equation for the initial spiral angle,α, the height f,r_1,ΔL,n, and s, variables; Likewise, with the variable values of the Trigonometric Partitions. Generalizing the equation with c to increase the size of the graph and with g to move it to any point on the x-axis then;

$r_3(p) = c * \sqrt{[\Delta L * e^{2(p-1)*\sin(\alpha/2^{(p-1)})-\alpha} + \sqrt{(r_1^2 + f^2)}]^2 - f^2} + g$, Where,$\Delta r = 2 * r * \sin^2(\alpha/2)$. With these equations of spiral angles, we can study the equation of prime numbers from two different points of view; therefore, the spiral angles will be: $\alpha = k * \arcsin[f/\sqrt{(f^2 + n^{2s})}]$, Additionally,s=a+bi and Where **n** is the posterior prime number, $r_1$is the previous prime number, **k** is the turns or twists, **f** is the height on the **Y** axis, and $\Delta L = \Delta r = r_3 - r_1$:the difference of the subsequent prime number minus the previous one. Consequently, care must be taken with which formula to use for the spiral angle of the prime numbers equation, such as The Riemann Z Function.**k** It is what causes the equation to rotate, in the plane of complex numbers, as **c** and **f** make the graph grow.In this section r_3 is function of **p**; $r_3(p)$.

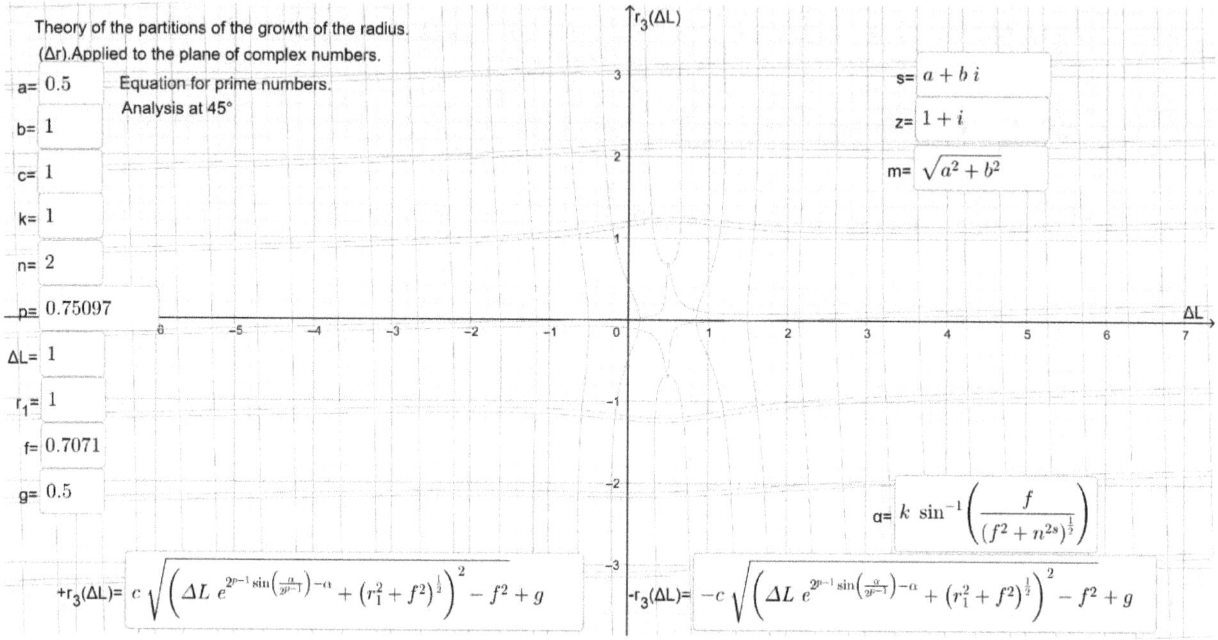

Figure 1.63: Graphics In The Plane Of Complex Numbers.

The behavior of the same equation in the plane of real numbers and the plane of complex numbers is different. In this graph two sliders **b** and **f** have been activated with a=0.5, therefore s=0.5+4.585i and g=0.5 moved the graph in the x-axis, the rest of the variables are fixed in the Cartesian plane to observe their behavior of the graph how it is growing and rotating in the Cartesian plane.

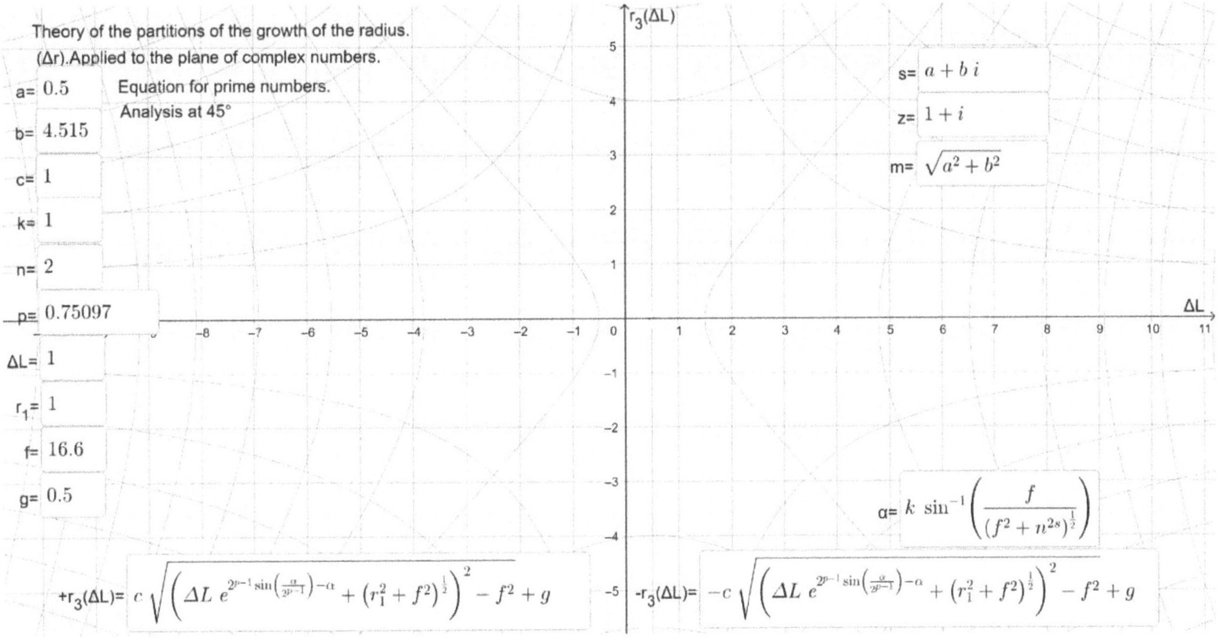

Figure 1.64: Graphics In The Plane Of Complex Numbers.

If we look at the pattern that makes $\alpha = k * \arcsin[f/\sqrt{(f^2 + n^{2s})}]$.With **f** and **b** varying in time, the equation is always going to take different paths, and it doesn't seem to have a periodic pattern. Even when the values of these variables **b** and **f** are bounded. The following graph is the spiral angle pattern $\alpha = k * \arcsin[f/\sqrt{(f^2 + n^{2s})}]$.With **b** and **f** variables, after a certain time, we will have the following graph, which follows a non-repetitive or non periodic pattern.

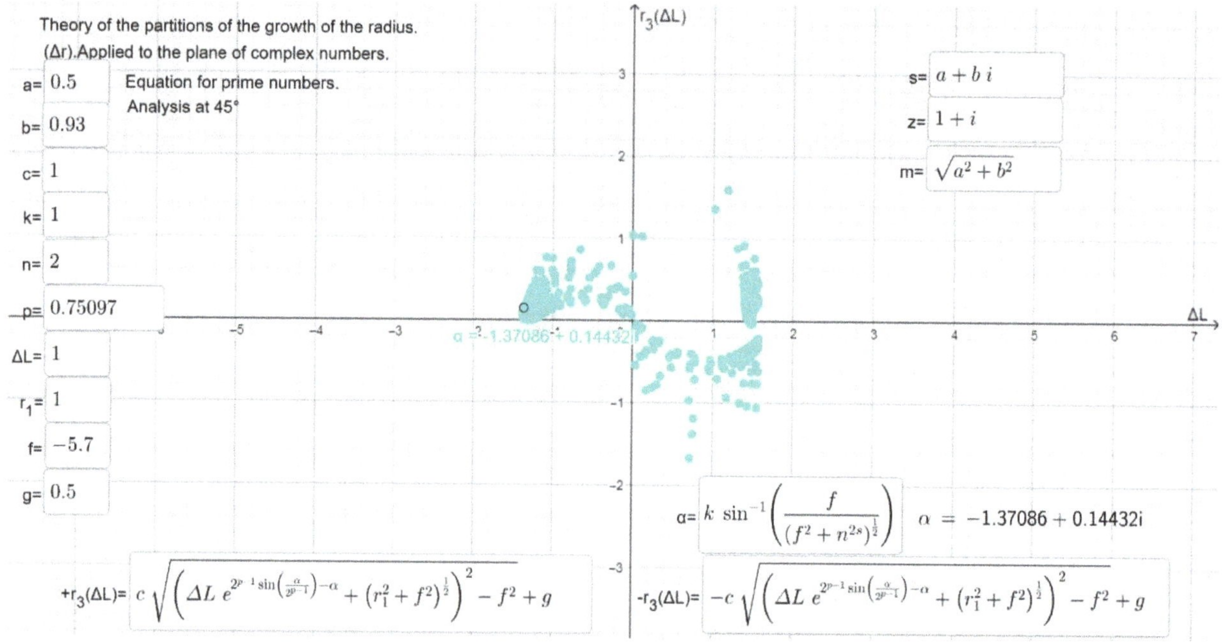

Figure 1.65: Graphics In The Plane Of Complex Numbers.

In the same way, if we consider the complex numbers Z_2 and Z_3 of the positive and negative equation we will obtain the following pattern after a certain time with the variables **b** and **f** varying in time.

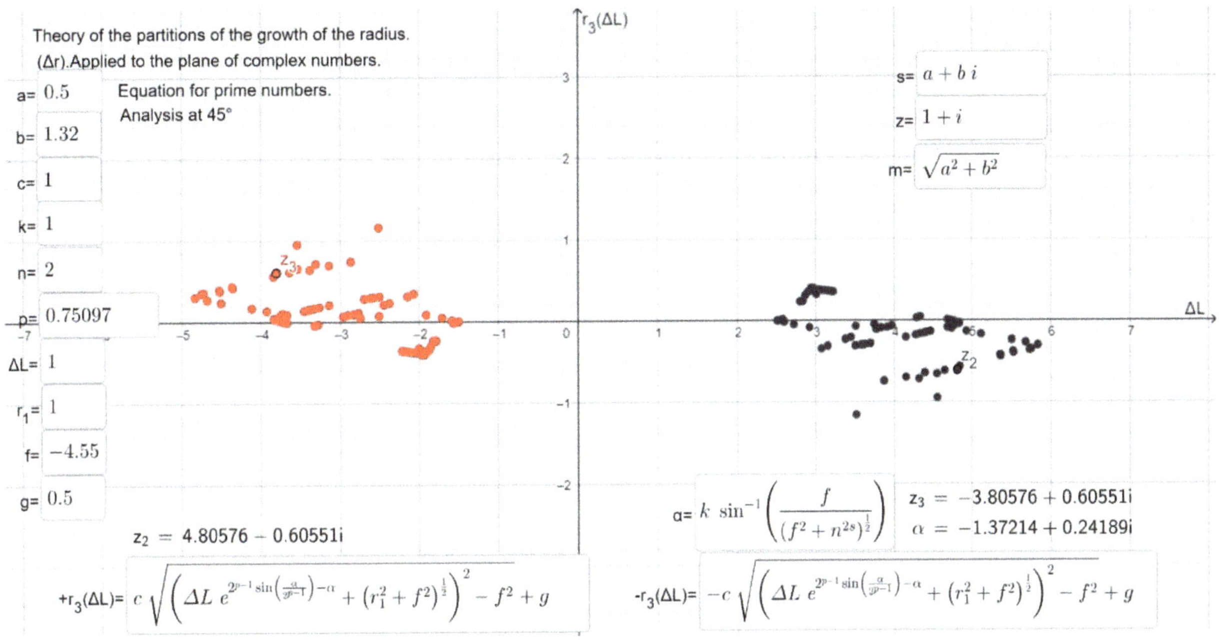

Figure 1.66: Graphics In The Plane Of Complex Numbers.

If we use the same equations with an initial analysis of 60°; in the complex plane with GeoGebra and its equation for the initial spiral angle, α, the height f,r_1, ΔL,**n**, and **s**, variables; Likewise, with the variable values of the Trigonometric Partitions. Generalizing the equation with **c** to increase the size of the graph and with **g** to move it to any point on the x-axis then; $r_3(p) = c * \sqrt{[\Delta L * e^{2^{(p-1)}*\sin(\alpha/2^{(p-1)})-\alpha} + \sqrt{(r_1^2 + f^2)}]^2 - f^2} + g$, Where,$\Delta r = 2 * r * \sin^2(\alpha/2)$.With these equations of spiral angles, we can study the equation of prime numbers from two different points of view; therefore, the spiral angles will be: $\alpha = 2 * k * \arcsin\sqrt{[\Delta L/(2 * n^s)]}$,Additionally, s=a+bi and $\zeta = k * \arcsin[f/\sqrt{(f^2 + n^{2s})}]$ Where **n** is the subsequent prime, r_1 is the previous prime number, **k** is the turns or twists, **f** is the height on the **Y** axis, and $\Delta L = \Delta r - r_3 - r_1$: the difference of the subsequent prime number minus the previous one. Consequently, care must be taken with which formula to use for the spiral angle of the prime numbers equation, such as The Riemann Z Function. The exponent **s** is what causes the equation to rotate, in the plane of complex numbers, just as **c** and **f** make the graph grow.

The behavior of the same equation in the plane of real numbers and in the plane of complex numbers is different. In this graph two sliders **b** and **f** have been activated with a=0.5, therefore s=0.5+2.935i and g=0.5 moved the graph in the x axis, the rest of the variables are fixed in the Cartesian plane to observe their behavior of the graph how it is growing and rotating in the Cartesian plane.

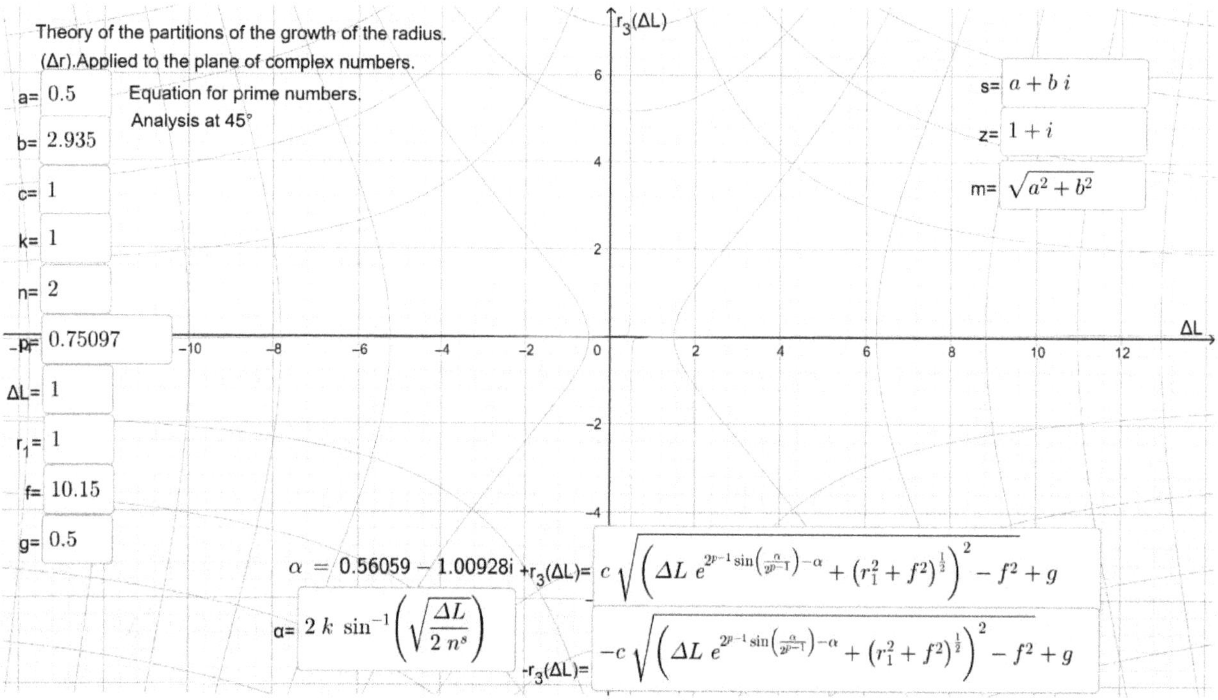

Figure 1.67: Graphics In The Plane Of Complex Numbers.

If we use the same equations with an initial analysis of 90°; in the complex plane with GeoGebra and its equation for the initial spiral angle,α, the height f=0,r_1,ΔL,**n**, and **s**, variables; Likewise, with the variable values of the Trigonometric Partitions. Generalizing the equation with **c** to increase the size of the graph and with **g** to move it to any point on the x-axis then; $r_3(p) = c * \Delta L * e^{[2^{(p-1)} * \sin(\alpha/2^{(p-1)}) - \alpha]} + r_1 + g$,where ,$\Delta r = 2 * r * \sin^2(\alpha/2)$.

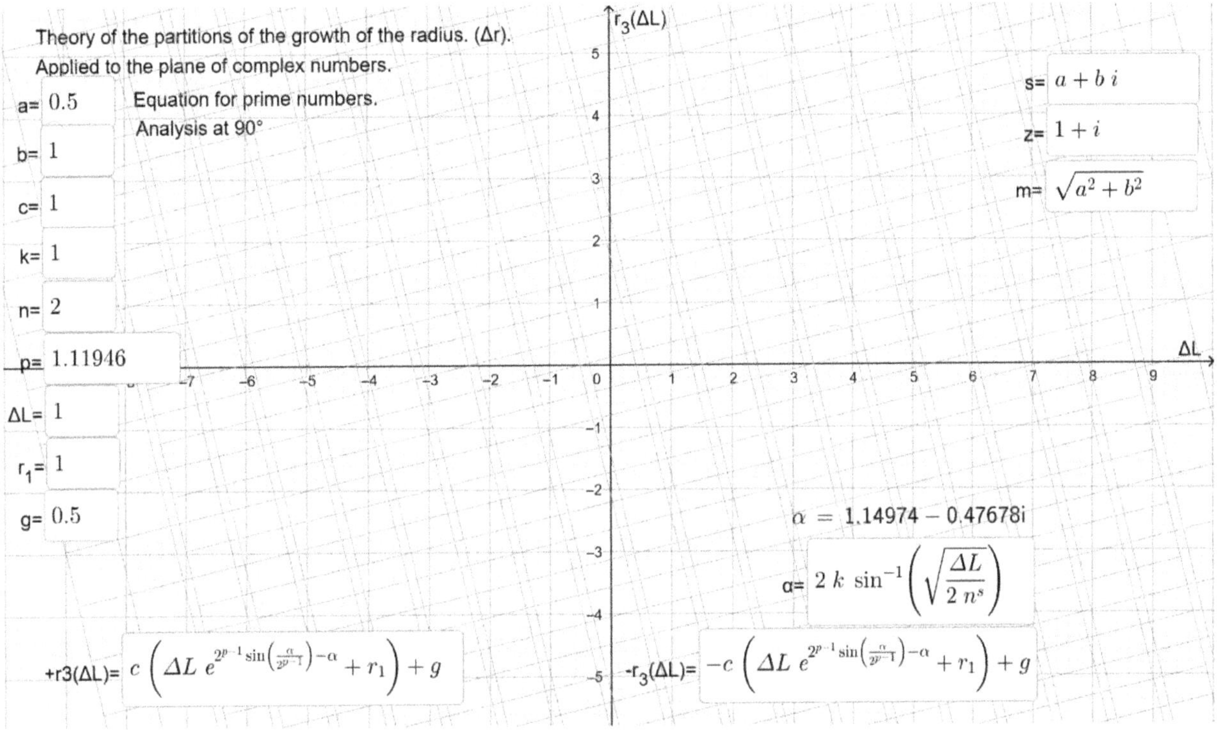

Figure 1.68: Graphics In The Plane Of Complex Numbers.

If we look at the pattern that makes $\alpha = 2 * k * \arcsin\left[\Delta L/\sqrt{(2 * n^s)}\right]$.With **f** and **b** varying in time, the equation is always going to take different paths, and appears to have a periodic pattern. Even when the values of these variables **b** and **f** are bounded. The following graph is the spiral angle pattern $\alpha = 2 * k * \arcsin\left[\Delta L/\sqrt{(2 * n^s)}\right]$, With **b** and **f** variables, after a certain time we will have the following graph which follows a repetitive and periodic pattern. In the same way if we consider the complex numbers of the positive and negative equation we will obtain the following pattern after a certain time with the variables **b** and **f** varying in time.

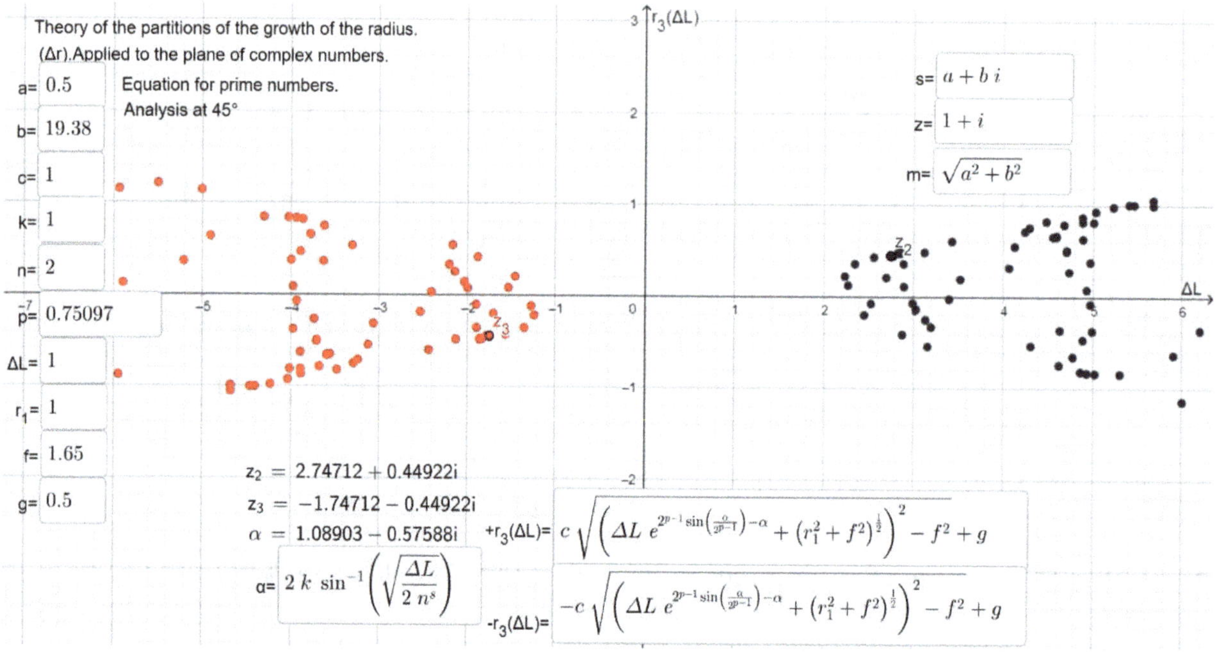

Figure 1.69: Graphics In The Plane Of Complex Numbers.

Chapter 2

Riemann Hypothesis Equation.

2.1 The Equation Of The Riemann Zeta Function In The Plane Of Real Numbers.

Riemann gave a formula for the number of primes less than **x** in terms the integral of $\frac{1}{\log(x)}$ and the roots (zeros) of the zeta function, defined by $\zeta(s) = 1 + \frac{1}{2^s} + \frac{1}{3^s} + \frac{1}{4^s}$+....He also formulated a conjecture about the location of these zeros, which fall into two classes: the "obvious zeros" -2, -4, -6, etc., and those whose real part lies between 0 and 1. Riemann's conjecture was that the real part of the non-obvious zeros is exactly $\frac{1}{2}$.That is, they all lie on a specific vertical line in the complex plane. Riemann checked the first few zeros of the zeta function by hand. They satisfy his hypothesis. By now over 1.5 billion zeros have been checked by computer. Very strong experimental evidence. But in mathematics we require a proof. A proof gives certainty, but, just as important, it gives understanding: it helps us understand why a result is true. Why is the Riemann hypothesis interesting? The closer the real part of the zeros lies to $\frac{1}{2}$,the more regular the distribution of the primes. To draw a statistical analogy, if the prime number theorem tells us something about the average distribution of the primes along the number line, then the Riemann hypothesis tells us something about the deviation from the average. The Clay Mathematics Institute, March 31th, 2023,

Riemann's 1859 manuscript,(Clay Mathematics Institute, n.d.-b)

https://www.claymath.org/publications/riemanns-1859-manuscript.

With the following schematic we can deduce various equations to obtain the formula of the Riemann Hypothesis in the real and imaginary plane. This same methodology can be applied to any initial angle that you want to develop an equation for The Riemann Z Function.

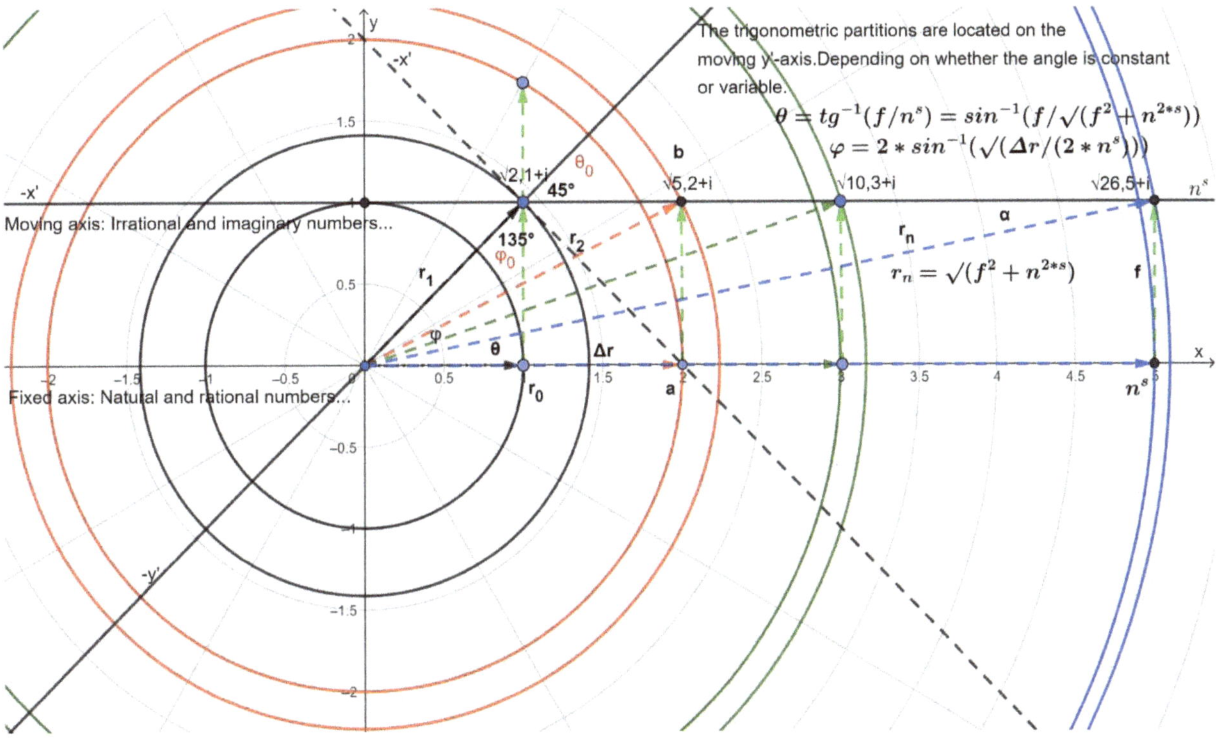

Figure 2.1: Riemann Z Function In The Plane Of Real Numbers.

Analysis of the Riemann Z Function with a constant initial angle of 45°. Applying the Law of Cosines we will obtain the following mathematical equation: $\theta_0 = 45° = \pi/4$; which is marked by the blue axis and is constant. $\varphi_0 = 135°$; Which is the constant or variable supplementary angle. $\Delta r = \Delta L$; which is the length module of the prime numbers and is variable.

$r_1 = \sqrt{f^2 + n^{(2*s)}}$, It is the first radius; which can be constant or variable. $r_2 = r_n$;which is the final radius or nth radius and is variable. $r_2^2 = r_1^2 + \Delta r^2 - 2*r_1*\Delta r*\cos(\varphi_0)$, Substituting into this equation we get: $r_2^2 = r_1^2 + \Delta L^2 - 2r_1 * \Delta L * \cos(135°) = r_1^2 + \Delta L^2 - 2r_1 * \Delta L *$

$(-\sqrt{2}/2) = r_1^2 + \Delta L^2 + \sqrt{2} * r_1 \Delta L$. Using the equation of the Trigonometric Partitions of the Chord. $r_2 = r_1 * e^{2^{(p+2)} * \sin^2(\theta/2^{p+1})}$; Squaring this equation we get:$r_2^2 = r_1^2 * e^{2^{(p+3)} * \sin^2(\theta/2^{p+1})}$, which is the same as: $r_n^2 = r_1^2 * e^{2^{(p+3)} * \sin^2(\theta/2^{p+1})}$.

Substituting two r_2 into the radius sub, we obtain: $r_1^2 * e^{2^{(p+3)} * \sin^2(\theta/2^{p+1})} = r_1^2 + \Delta L^2 + \sqrt{2} * r_1 * \Delta L$, Solving.$-r_1^2 * e^{2^{(p+3)} * \sin^2(\theta/2^{p+1})} + r_1^2 + \Delta L^2 + \sqrt{2} * r_1 * \Delta L = 0$,

factoring.

$r_1^2 * (1 - e^{2^{(p+3)} * \sin^2(\theta/2^{p+1})}) + \Delta L^2 + \sqrt{2} * r_1 * \Delta L = 0$ ordering as a quadratic equation;

$r_1^2 * (1 - e^{2^{(p+3)} * \sin^2(\theta/2^{p+1})}) + \sqrt{2} * r_1 * \Delta L + \Delta L^2 = 0$,

applying the formula of the quadratic equation will obtain two solutions for r_1 ,one negative and one positive.

$r_{1a} = \sqrt{2} * \Delta L * [-1 - [1 - 2 * (1e^{2^{(p+3)} * \sin^2(\theta/2^{p+1})})]]/[2 * (1 - e^{2^{(p+3)} * \sin^2(\theta/2^{p+1})})]$.

$r_{1b} = \sqrt{2} * \Delta L * [-1 + [1 - 2 * (1 - e^{2^{(p+3)} * \sin^2(\theta/2^{p+1})})]]/[2 * (1 - e^{2^{(p+3)} * \sin^2(\theta/2^{p+1})})]$.

Substituting each of the solutions to: $r_1 = \sqrt{f^2 + n^{2*s}}$,

It is the first radius; that forms circular plane waves in a radial shape.

$\sqrt{(f^2 + n^{2*s})} = \sqrt{2} * \Delta L * [-1 - [1 - 2 * (1 - e^{2^{(p+3)} * \sin^2(\theta/2^{p+1})})]]/[2 * (1 - e^{2^{(p+3)} * \sin^2(\theta/2^{p+1})})]$.

Finally,solve for n^s; which represents the plane circular waves that form radii in the horizontal x-axis. Likewise,two solutions will be obtained,one negative and one positive of the square roost's.

$n^s = \sqrt{[\Delta L^2 * [-1 - \sqrt{[1 - 2 * (1 - e^{2^{(p+3)} * \sin^2(\theta/2^{p+1})})]]}^2 - 2 * f^2 * (1 - e^{2^{(p+3)} * \sin^2(\theta/2^{p+1})})^2]}$
$/\sqrt{2} * (1 - e^{2^{(p+3)} * \sin^2(\theta/2^{p+1})})$.

Inverting n^s to obtain the summation;

$1/n^s = \sqrt{2} * (1 - e^{2^{(p+3)} * \sin^2(\theta/2^{p+1})})$
$/\sqrt{[\Delta L^2 * [-1 - \sqrt{[1 - 2 * (1 - e^{2^{(p+3)} * \sin^2(\theta/2^{p+1})})]]}^2 - 2 * f^2 * (1 - e^{2^{(p+3)} * \sin^2(\theta/2^{p+1})})^2]}$

Therefore the first negative solution of the quadratic equation will be: From **n** equal to one to infinity.

In this equation, there will be four solutions, the first two of the quadratic solution and the other two of the solutions of the square roots.

$Z(p)_1 = \sum_{n=1}^{\infty}(1/n^s) = \sqrt{2} * (1 - e^{2^{(p+3)} * \sin^2(\theta/2^{p+1})})$
$/\sqrt{[\Delta L^2 * [-1 - \sqrt{[1 - 2 * (1 - e^{2^{(p+3)} * \sin^2(\theta/2^{p+1})})]]}^2 - 2 * f^2 * (1 - e^{2^{(p+3)} * \sin^2(\theta/2^{p+1})})^2]}$.

$Z(p)_2 = \sum_{n=1}^{\infty}(1/n^s) = -\sqrt{2} * (1 - e^{2^{(p+3)} * \sin^2(\theta/2^{p+1})})$
$/\sqrt{[\Delta L^2 * [-1 - \sqrt{[1 - 2 * (1 - e^{2^{(p+3)} * \sin^2(\theta/2^{p+1})})]]}^2 - 2 * f^2 * (1 - e^{2^{(p+3)} * \sin^2(\theta/2^{p+1})})^2]}$.

Similarly, the second positive solution of the quadratic equation will be:

$Z(p)_3 = \sum_{n=1}^{\infty}(1/n^s) = \sqrt{2} * (1 - e^{2^{(p+3)} * \sin^2(\theta/2^{p+1})})$
$/\sqrt{[\Delta L^2 * [-1 + \sqrt{[1 - 2 * (1 - e^{2^{(p+3)} * \sin^2(\theta/2^{p+1})})]]}^2 - 2 * f^2 * (1 - e^{2^{(p+3)} * \sin^2(\theta/2^{p+1})})^2]}$.

$Z(p)_4 = \sum_{n=1}^{\infty}(1/n^s) = -\sqrt{2} * (1 - e^{2^{(p+3)} * \sin^2(\theta/2^{p+1})})$
$/\sqrt{[\Delta L^2 * [-1 - \sqrt{[1 - 2 * (1 - e^{2^{(p+3)} * \sin^2(\theta/2^{p+1})})]]}^2 - 2 * f^2 * (1 - e^{2^{(p+3)} * \sin^2(\theta/2^{p+1})})^2]}$.

Finally, this equation can be used with a constant spiral angle to observe its behavior just like the Prime Numbers equation. $\theta = 45° = \pi/4$, Substituting in $\theta/2^{(p+1)}$, Therefore the spiral angle, with **k** turns, will be: $\pi * k/2^{(p+3)}$

$Z(p)_1 = \sum_{n=1}^{\infty}(1/n^s) = \sqrt{2} * (1 - e^{2^{(p+3)} * \sin^2(\pi * k/2^{p+3})})$
$/\sqrt{[\Delta L^2 * [-1 - \sqrt{[1 - 2 * (1 - e^{2^{(p+3)} * \sin^2(\pi * k/2^{p+3})})]]}^2 - 2 * f^2 * (1 - e^{2^{(p+3)} * \sin^2(\pi * k/2^{p+3})})^2]}$.

$Z(p)_3 = \sum_{n=1}^{\infty}(1/n^s) = \sqrt{2} * (1 - e^{2^{(p+3)} * \sin^2(\pi * k/2^{p+3})})$
$/\sqrt{[\Delta L^2 * [-1 + \sqrt{[1 - 2 * (1 - e^{2^{(p+3)} * \sin^2(\pi * k/2^{p+3})})]]}^2 - 2 * f^2 * (1 - e^{2^{(p+3)} * \sin^2(\pi * k/2^{p+3})})^2]}$.

$Z(p)_{1x} = \cos(\pi * k/4) * \sqrt{2} * (1 - e^{2^{(p+3)} * \sin^2(\pi * k/2^{p+3})})$
$/\sqrt{[\Delta L^2 * [-1 - \sqrt{[1 - 2 * (1 - e^{2^{(p+3)} * \sin^2(\pi * k/2^{p+3})})]]}^2 - 2 * f^2 * (1 - e^{2^{(p+3)} * \sin^2(\pi * k/2^{p+3})})^2]}$.

$Z(p)_{1y} = \sin(\pi * k/4) * \sqrt{2} * (1 - e^{2^{(p+3)} * \sin^2(\pi * k/2^{p+3})})$
$/\sqrt{[\Delta L^2 * [-1 - \sqrt{[1 - 2 * (1 - e^{2^{(p+3)} * \sin^2(\pi * k/2^{p+3})})]]}^2 - 2 * f^2 * (1 - e^{2^{(p+3)} * \sin^2(\pi * k/2^{p+3})})^2]}$.

$Z(p)_{3x} = \cos(\pi * k/4) * \sqrt{2} * (1 - e^{2^{(p+3)} * \sin^2(\pi * k/2^{p+3})})$
$/\sqrt{[\Delta L^2 * [-1 + \sqrt{[1 - 2 * (1 - e^{2^{(p+3)} * \sin^2(\pi * k/2^{p+3})})]]}^2 - 2 * f^2 * (1 - e^{2^{(p+3)} * \sin^2(\pi * k/2^{p+3})})^2]}$.

$Z(p)_{3y} = \sin(\pi * k/4) * \sqrt{2} * (1 - e^{2^{(p+3)} * \sin^2(\pi * k/2^{p+3})})$
$/\sqrt{[\Delta L^2 * [-1 + \sqrt{[1 - 2 * (1 - e^{2^{(p+3)} * \sin^2(\pi * k/2^{p+3})})]]}^2 - 2 * f^2 * (1 - e^{2^{(p+3)} * \sin^2(\pi * k/2^{p+3})})^2]}$.

With these equations, the trivial zeros can be obtained, in the plane of the real numbers in the negative X-axis. If **k** turns, **p** trigonometric partitions, ΔL the length module of the prime numbers, **n** the subsequent real prime number, are varying in time; Without these variables being bounded, the graph +Z(p) and -Z(p) will be passing through the values: -1,-2,-3.-4, and so on. It is worth mentioning that this initial angle can be any value; $\theta_0 = 0°, 30°, 45°, 90°$....When it is ninety degrees the equation is simpler. Something interesting happens with this 90° analysis equation: the graph that is generated is in one piece, not two parts as in the rest of the equations. The rest of the equations generate graphs in two parts, one with the positive equation and another with the negative equation.

$1/n^s = \sqrt{[e^{[2^{(p+3)}*\sin^2(\theta/2^{(p+1)})]} - 1]/\Delta L}.$

The next graph is the equation

$$Z(p)_2 = -\sqrt{2}*(1-e^{2^{(p+3)}*\sin^2(\pi*k/2^{p+3})})/ \sqrt{[\Delta L^2*[-1-\sqrt{[1-2*(1-e^{2^{(p+3)}*\sin^2(\pi*k/2^{p+3})})]]}]^2 - 2*f^2*(1-e^{2^{(p+3)}*\sin^2(\pi*k/2^{p+3})})^2]}.$$

Which is a solution in the real number plane. In the 2D plane the equation generates dots and for it to be visible you have to select mark the trace in GeoGebra. In the 3D plane, the equation generates arcs or circles depending on the values of **k** turns, and the initial spiral angle selected for its study of the Riemann Z Function. Where **p** are the trigonometric partitions, **n** is the posterior prime number, ΔL is the growth of the radius of the posterior prime numbers minus the previous one, **s** is the exponent of the real numbers, **f** is the height on the **Y** axis, for the starting point of the initial analysis with the theta spiral angle θ; in this case at 45°.

In this study **n** and ΔL must be different from zero.

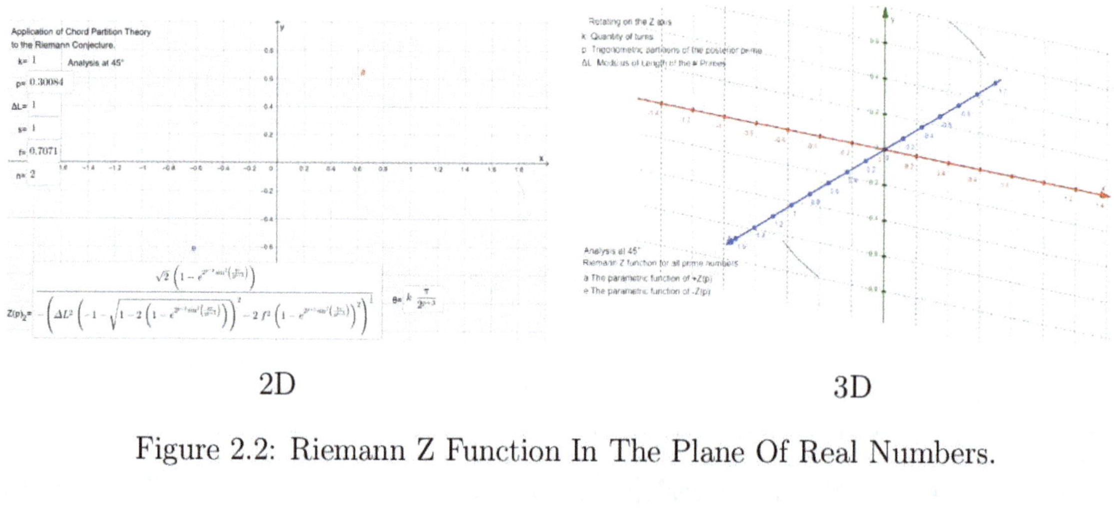

2D 3D

Figure 2.2: Riemann Z Function In The Plane Of Real Numbers.

The following graph is of the positive and negative equation $Z(p)_2$;with its **X** and **Y** components.

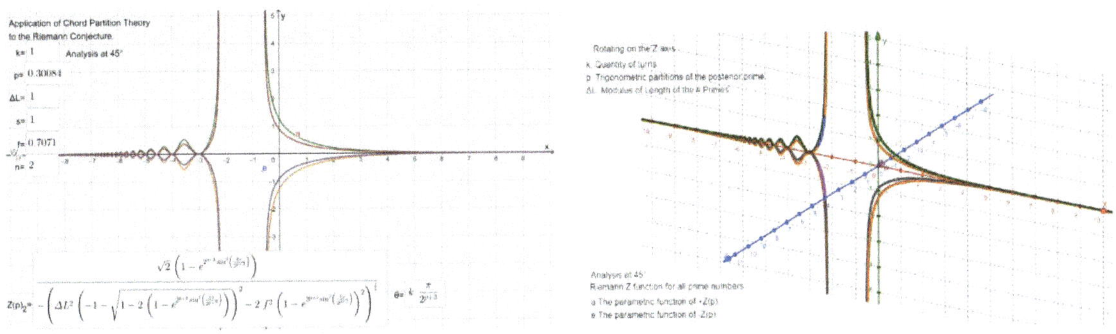

2D 3D

Figure 2.3: Riemann Z Function In The Plane Of Real Numbers.

The following graph is of the positive and negative equation $Z(p)_2$;with its **X** and **Y** components. With **k** turns and the height **f**, varying in time.

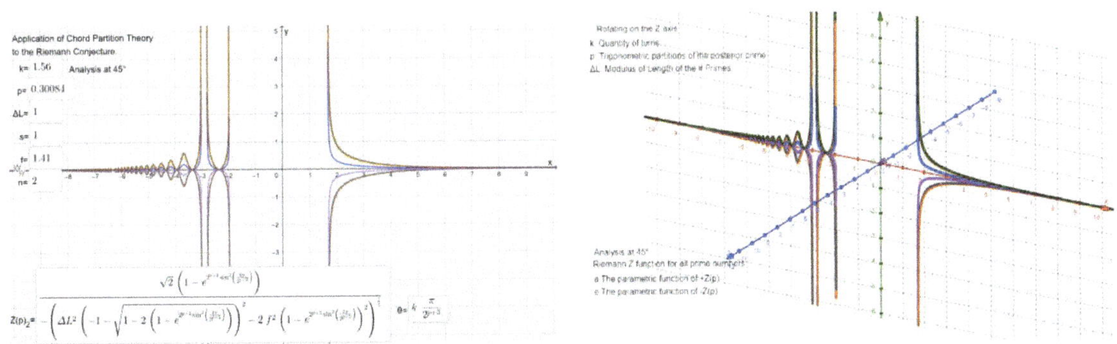

2D 3D

Figure 2.4: Riemann Z Function In The Plane Of Real Numbers.

With the following schematic, we can deduce various equations to obtain the formula of the Riemann Hypothesis in the real and imaginary plane. This same methodology can be applied to any initial angle that you want to develop an equation for The Riemann Z Function. This type of analysis is different from the Prime Number equations.

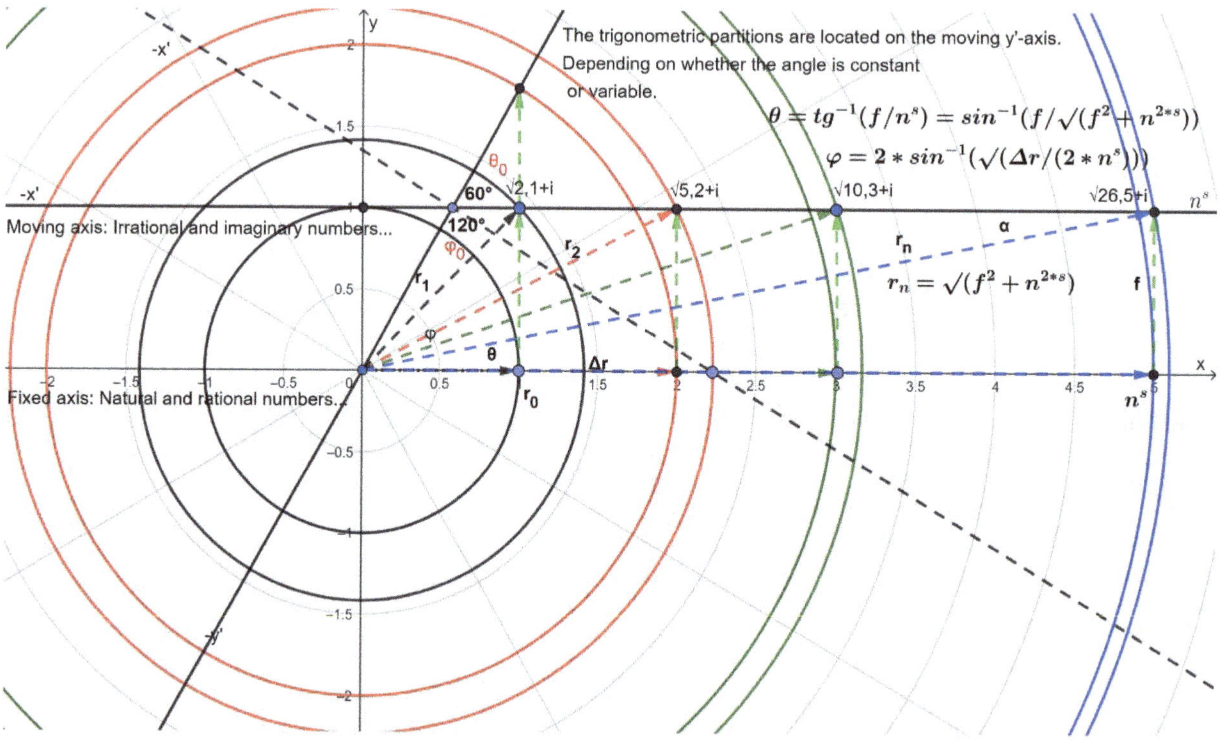

Figure 2.5: Riemann Z Function In The Plane Of Real Numbers.

Analysis of the Riemann Z Function with a constant initial angle of 60°. Applying the Law of Cosines we will obtain the following mathematical equation: $\theta_0 = 60° = \pi/3$; which is marked by the blue axis and is constant. $\varphi_0 = 120°$; Which is the constant or variable supplementary angle. $\Delta r = \Delta L$; which is the length module of the prime numbers and is variable.

$r_1 = \sqrt{f^2 + n^{2s}}$, It is the first radius; which can be constant and variable. $r_2 = r_n$; which is the final radius or nth radius and is variable. $r_2^2 = r_1^2 + \Delta r^2 - 2r_1 * \Delta r * \cos(\varphi_0)$,

Substituting into this equation we get:

$r_2^2 = r_1^2 + \Delta L^2 - 2r_1 * \Delta L * \cos(120°) = r_1^2 + \Delta L^2 - 2r_1 * \Delta L * (-1/2) = r_1^2 + \Delta L^2 + r_1 * \Delta L.$

Using the equation of the Trigonometric Partitions of the Chord.

$r_2 = r_1 * e^{2(p+2)*\sin^2(\theta/2^{p+1})}$; Squaring this equation we get:

$r_2^2 = r_1^2 * e^{2(p+3)*\sin^2(\theta/2^{p+1})}$, which is the same as:

$r_n^2 = r_1^2 * e^{2(p+3)*\sin^2(\theta/2^{p+1})}$. Substituting two r_2 into the radius sub, we get:

$r_1^2 * e^{2(p+3)*\sin^2(\theta/2^{p+1})} = r_1^2 + \Delta L^2 + r_1 * \Delta L$, solving.

$-r_1^2 * e^{2(p+3)*\sin^2(\theta/2^{p+1})} + r_1^2 + \Delta L^2 + r_1 * \Delta L = 0$, Factoring.

$r_1^2 * [1 - e^{2(p+3)*\sin^2(\theta/2^{p+1})}] + \Delta L^2 + r_1 * \Delta L = 0$,

ordering as a quadratic equation;

$r_1^2 * [1 - e^{2(p+3)*\sin^2(\theta/2^{p+1})}] + r_1 * \Delta L + \Delta L^2 = 0$,

Applying the formula of the quadratic equation, two solutions for r_1 will be obtained:

$r_{1a} = \Delta L * [-1 - [1 - 4 * (1 - e^{2(p+3)*\sin^2(\theta/2^{p+1})})]]/[2 * (1 - e^{2(p+3)*\sin^2(\theta/2^{p+1})})]$

$r_{1b} = \Delta L * [-1 + [1 - 4 * (1 - e^{2(p+3)*\sin^2(\theta/2^{p+1})})]]/[2 * (1 - e^{2(p+3)*\sin^2(\theta/2^{p+1})})]$

Substituting in each of the solutions to: $r_1 = \sqrt{f^2 + n^{(2*s)}}$, It is the first radius; that forms circular plane waves in a radial shape.

$\sqrt{f^2 + n^{(2*s)}} = \Delta L * [-1 - [1 - 4 * (1 - e^{2(p+3)*\sin^2(\theta/2^{p+1})})]]/[2 * (1 - e^{2(p+3)*\sin^2(\theta/2^{p+1})})]$

Finally, solving for n^s; which represents the plane circular waves that form radii in the horizontal x-axis.Likewise, two solutions will be obtained, one negative and one positive of the square roots.

$n^s = \sqrt{[\Delta L^2 * [-1 - \sqrt{[1 - 4 * (1 - e^{2(p+3)*\sin^2(\theta/2^{p+1})})}]]^2 - 4f^2 * (1 - e^{2(p+3)*\sin^2(\theta/2^{p+1})})]}$
$/[2 * (1 - e^{2(p+3)*\sin^2(\theta/2^{p+1})})]$,

Inverting n^s to obtain the summation; $(1/n^s) = [2 * (1 - e^{2(p+3)*\sin^2(\theta/2^{p+1})})]/$
$\sqrt{[\Delta L^2 * [-1 - \sqrt{[1 - 4 * (1 - e^{2(p+3)*\sin^2(\theta/2^{p+1})})}]]^2 - 4f^2 * (1 - e^{2(p+3)*\sin^2(\theta/2^{p+1})})]}$,

Therefore the first negative solution of the quadratic equation will be: From **n** equal to one to infinity.

In this equation, there will be 4 solutions, the first two of the quadratic solution and the other two of the solutions of the square roots.

$$Z(p)_1 = \sum_{n=1}^{\infty}(1/n^s) = [2*(1-e^{2(p+3)*\sin^2(\theta/2^{p+1})})]/\sqrt{[\Delta L^2*[-1-\sqrt{[1-4*(1-e^{2(p+3)*\sin^2(\theta/2^{p+1})})]]^2}-4f^2*(1-e^{2(p+3)*\sin^2(\theta/2^{p+1})})]}.$$

$$Z(p)_2 = \sum_{n=1}^{\infty}(1/n^s) = -[2*(1-e^{2(p+3)*\sin^2(\theta/2^{p+1})})]/\sqrt{[\Delta L^2*[-1-\sqrt{[1-4*(1-e^{2(p+3)*\sin^2(\theta/2^{p+1})})]]^2}-4f^2*(1-e^{2(p+3)*\sin^2(\theta/2^{p+1})})]}.$$

Similarly, the second positive solution of the quadratic equation will be:

$$Z(p)_3 = \sum_{n=1}^{\infty}(1/n^s) = [2*(1-e^{2(p+3)*\sin^2(\theta/2^{p+1})})]/\sqrt{[\Delta L^2*[-1+\sqrt{[1-4*(1-e^{2(p+3)*\sin^2(\theta/2^{p+1})})]]^2}-4f^2*(1-e^{2(p+3)*\sin^2(\theta/2^{p+1})})]}.$$

$$Z(p)_4 = \sum_{n=1}^{\infty}(1/n^s) = -[2*(1-e^{2(p+3)*\sin^2(\theta/2^{p+1})})]/\sqrt{[\Delta L^2*[-1+\sqrt{[1-4*(1-e^{2(p+3)*\sin^2(\theta/2^{p+1})})]]^2}-4f^2*(1-e^{2(p+3)*\sin^2(\theta/2^{p+1})})]}.$$

Finally, this equation can be used with a constant spiral angle to observe its behavior just like the prime numbers equation. $\theta = 60° = \pi/3$, Substituting in $\theta/2^{(p+1)}$, Therefore the spiral angle, with **k** turns, will be: $\pi * k/(3*2^{(p+1)})$

$$Z(p)_1 = \sum_{n=1}^{\infty}(1/n^s) = [2*(1-e^{2(p+3)*\sin^2(\pi*k/(3*2^{p+1}))})]/\sqrt{[\Delta L^2*[-1-\sqrt{[1-4*(1-e^{2(p+3)*\sin^2(\pi*k/(3*2^{p+1}))})]]^2}-4f^2*(1-e^{2(p+3)*\sin^2(\pi*k/(3*2^{p+1}))})]}.$$

$$Z(p)_3 = \sum_{n=1}^{\infty}(1/n^s) = [2*(1-e^{2(p+3)*\sin^2(\pi*k/(3*2^{p+1}))})]/\sqrt{[\Delta L^2*[-1+\sqrt{[1-4*(1-e^{2(p+3)*\sin^2(\pi*k/(3*2^{p+1}))})]]^2}-4f^2*(1-e^{2(p+3)*\sin^2(\pi*k/(3*2^{p+1}))})]}.$$

The polar coordinates **X** and **Y** will be:

$$Z(p)_{(1x)} = \cos(\pi*k/3)*[2*(1-e^{2(p+3)*\sin^2(\pi*k/(3*2^{p+1}))})]/\sqrt{[\Delta L^2*[-1-\sqrt{[1-4*(1-e^{2(p+3)*\sin^2(\pi*k/(3*2^{p+1}))})]]^2}-4f^2*(1-e^{2(p+3)*\sin^2(\pi*k/(3*2^{p+1}))})]}.$$

$$Z(p)_{(1y)} = \sin(\pi*k/3)*[2*(1-e^{2(p+3)*\sin^2(\pi*k/(3*2^{p+1}))})]/\sqrt{[\Delta L^2*[-1-\sqrt{[1-4*(1-e^{2(p+3)*\sin^2(\pi*k/(3*2^{p+1}))})]]^2}-4f^2*(1-e^{2(p+3)*\sin^2(\pi*k/(3*2^{p+1}))})]}.$$

$$Z(p)_{(3x)} = \cos(\pi*k/3)*[2*(1-e^{2(p+3)*\sin^2(\pi*k/(3*2^{p+1}))})]/\sqrt{[\Delta L^2*[-1+\sqrt{[1-4*(1-e^{2(p+3)*\sin^2(\pi*k/(3*2^{p+1}))})]]^2}-4f^2*(1-e^{2(p+3)*\sin^2(\pi*k/(3*2^{p+1}))})]}.$$

$$Z(p)_{(3y)} = \sin(\pi*k/3)*[2*(1-e^{2(p+3)*\sin^2(\pi*k/(3*2^{p+1}))})]/\sqrt{[\Delta L^2*[-1+\sqrt{[1-4*(1-e^{2(p+3)*\sin^2(\pi*k/(3*2^{p+1}))})]]^2}-4f^2*(1-e^{2(p+3)*\sin^2(\pi*k/(3*2^{p+1}))})]}.$$

With these equations, the trivial zeros can be obtained, in the plane of the real numbers in the negative X-axis. If **k** turns, **p** trigonometric partitions, ΔL the length module of the prime numbers, **n** the subsequent real prime number, are varying in time; Without these variables being bounded, the graph +Z(p) and -Z(p) will be passing through the values: -1,-2,-3.-4, and so on. The next graph is the equation;

$$Z(p)_2 = -[2 * (1 - e^{2(p+3)*\sin^2(\pi*k/(3*2^{p+1})})] / \sqrt{[\Delta L^2 * [-1 - [1 - 4 * \sqrt{(1 - e^{2(p+3)*\sin^2(\pi*k/(3*2^{p+1}))})}]]^2 - 4f^2 * (1 - e^{2(p+3)*\sin^2(\pi*k/(3*2^{p+1}))})]}.$$

Which is a solution in the real number plane. In the 2D plane the equation generates dots and for it to be visible you must have to select mark the trace in GeoGebra. In the 3D plane the equation generates arcs or circles depending on the values of **k** turns, and the initial spiral angle selected for your study of the Riemann Z Function. Where **p** are the trigonometric partitions, **n** is the posterior prime number, ΔL is the growth of the radius of the posterior prime number minus the previous one, **s** is the exponent of the real numbers, **f** is the height on the **Y** axis for the point starting point of the initial analysis with the theta spiral angle θ ; in this case at 60°.

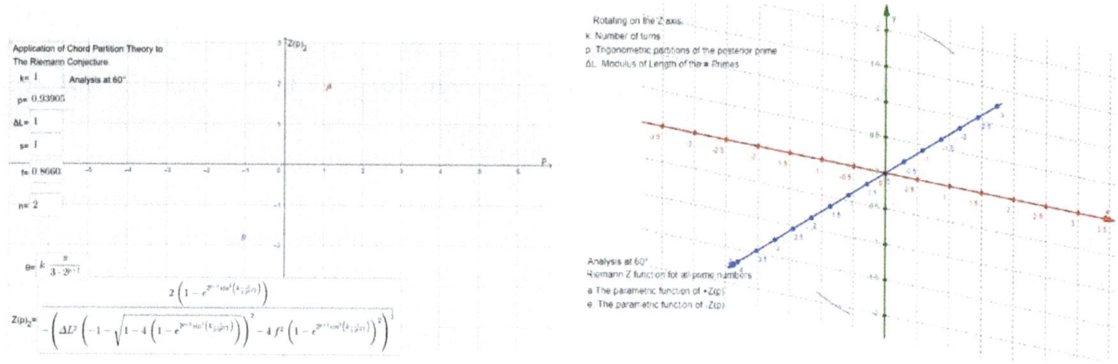

2D 3D

Figure 2.6: Riemann Z Function In The Plane Of Real Numbers.

The following graph is of the positive and negative equation $Z(p)_2$; with its **X** and **Y** components.

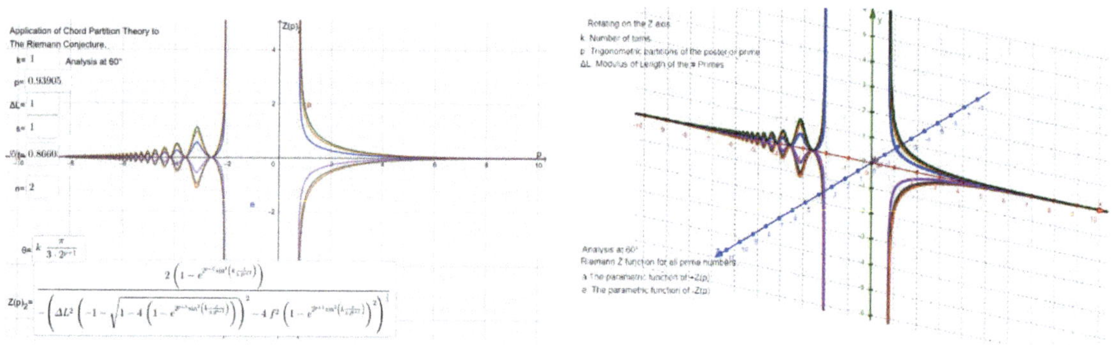

2D 3D

Figure 2.7: Riemann Z Function In The Plane Of Real Numbers.

The following graph is of the positive and negative equation $Z(p)_2$;with its **X** and **Y** components. With **k** turns and the height **f**, varying in time.

The following graph is the equation that can be used with all values varying in time:

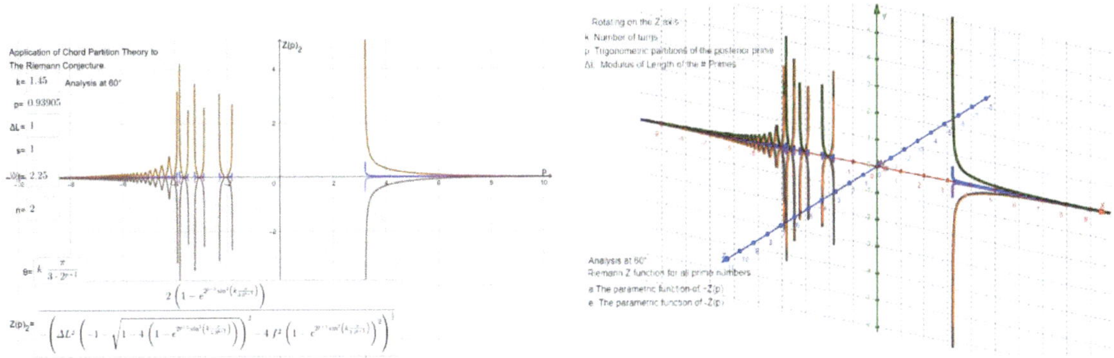

2D 3D

Figure 2.8: Riemann Z Function In The Plane Of Real Numbers.

$$Z(p)_2 = \sum_{n=1}^{\infty}(1/n^s) = -\sqrt{2} * (1 - e^{2(p+3)} * \sin^2(\theta/2^{p+1}))$$
$$/\sqrt{[\Delta L^2 * [-1 - \sqrt{[1 - 2 * (1 - e^{2(p+3)} * \sin^2(\theta/2^{p+1}))]]}^2 - 2 * f^2 * (1 - e^{2(p+3)} * \sin^2(\theta/2^{p+1}))^2]}.$$

With variable spiral angle;$\theta = k * \arcsin[f/\sqrt{(f^2 + n^{2s})}]$ or $\theta = -k * \arcsin[f/\sqrt{(f^2 + n^{2s})}]$. Which is a solution in the real number plane. In the 2D plane the equation generates dots and for it to be visible you have to select mark the trace in GeoGebra. In the 3D plane the equation generates arcs or circles depending on the values of **k** turns, and the initial spiral angle selected for your study of the Riemann Z Function. Where **p** are the trigonometric partitions, **n** is the posterior prime number, ΔL is the growth of the radius of the posterior prime number minus the previous one, **s** is the exponent of the real numbers, **f** is the height on the **Y** axis, for the starting point of the initial analysis with the theta spiral angle θ; in this case at 45°.

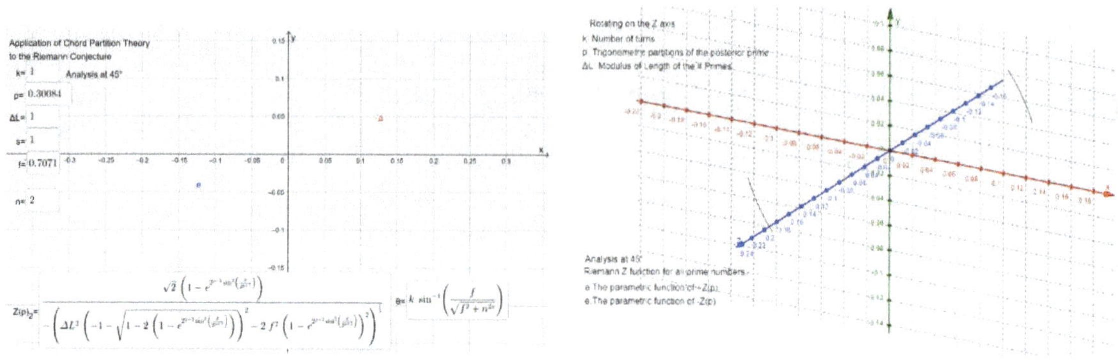

2D 3D

Figure 2.9: Riemann Z Function In The Plane Of Real Numbers.

The following graph is of the positive and negative equation $Z(p)_2$; with its X and Y components.

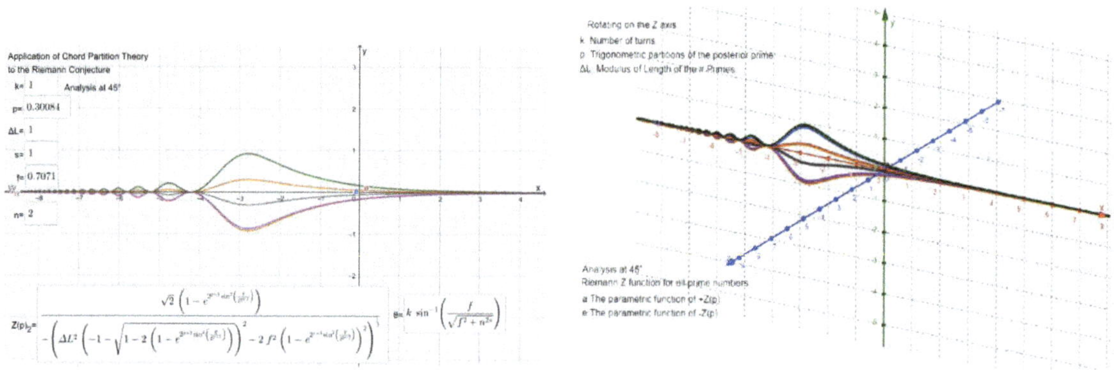

2D 3D

Figure 2.10: Riemann Z Function In The Plane Of Real Numbers.

The following graph is of the positive and negative equation $Z(p)_2$; with its **X** and **Y** components. With **k** turns, spiral angle ΔL and height **f**, varying in time. The rest of the constant variables.

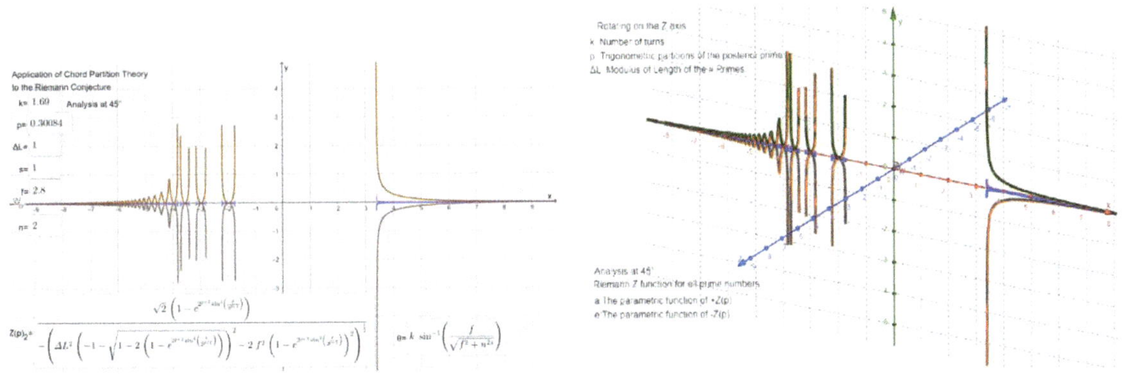

2D 3D

Figure 2.11: Riemann Z Function In The Plane Of Real Numbers.

The following graph is the equation that can be used with all values varying in time:
$Z(p)_2 = \sum_{n=1}^{\infty}(1/n^s) = -[2 * (1 - e^{2(p+3)*\sin^2(\theta/2^{p+1})})]$
$/\sqrt{[\Delta L^2 * [-1 - \sqrt{[1 - 4 * (1 - e^{2(p+3)*\sin^2(\theta/2^{p+1})})]}]^2 - 4f^2 * (1 - e^{2(p+3)*\sin^2(\theta/2^{p+1})})]}$.
With variable spiral angle;$\theta = k * \arcsin[f/\sqrt{(f^2 + n^{2*s})}]$ Which is a solution in the real number plane. In the 2D plane the equation generates dots and for it to be visible you have to select mark the trace in GeoGebra. In the 3D plane the equation generates arcs or circles depending on the values of **k** turns, and the initial spiral angle selected for your study of the Riemann Z Function. Where **p** are the trigonometric partitions, **n** is the posterior prime number, ΔL is the growth of the radius of the posterior prime number minus the previous one, **s** is the exponent of the real numbers, **f** is the height on the **Y** axis for the point starting point of the initial analysis with the theta spiral angle θ; in this case at 60°.

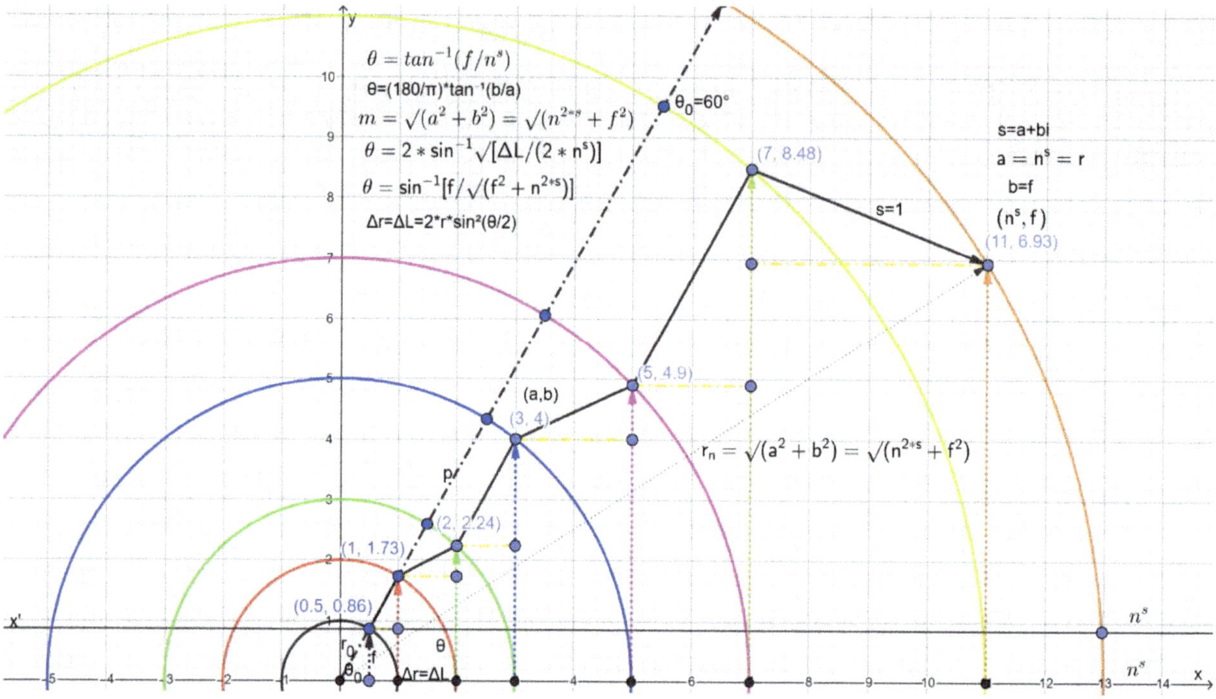

Figure 2.12: Circular wave graph of Prime Numbers.

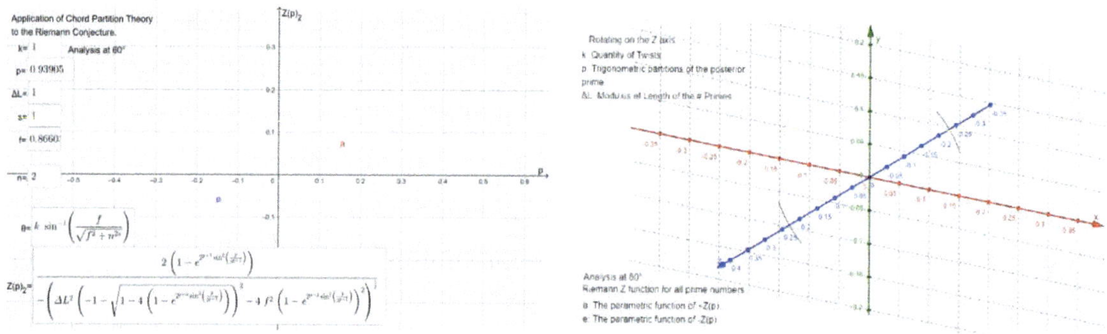

2D 3D

Figure 2.13: Riemann Z Function In The Plane Of Real Numbers.

The following graph is of the positive and negative equation $Z(p)_2$; with its **X** and **Y** components

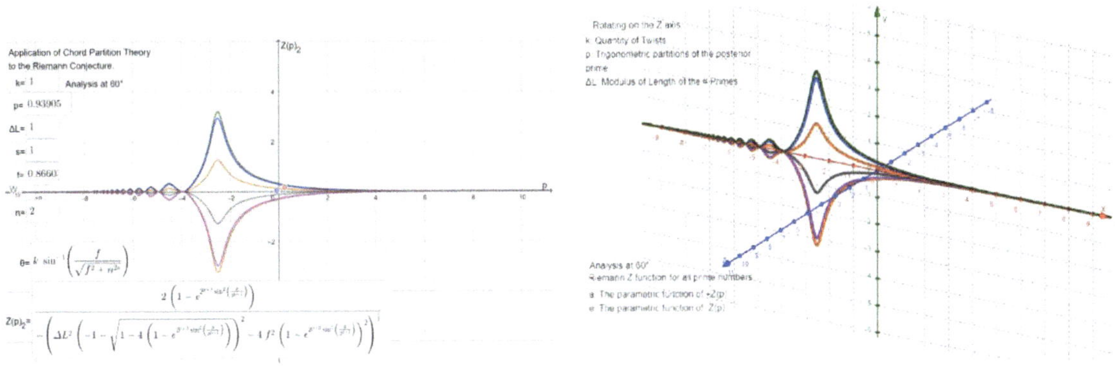

2D 3D

Figure 2.14: Riemann Z Function In The Plane Of Real Numbers.

The following graph is of the positive and negative equation $Z(p)_2$; with its **X** and **Y** components. With **k** turns, spiral angle θ and height **f**, varying in time. The rest of the constant variables.

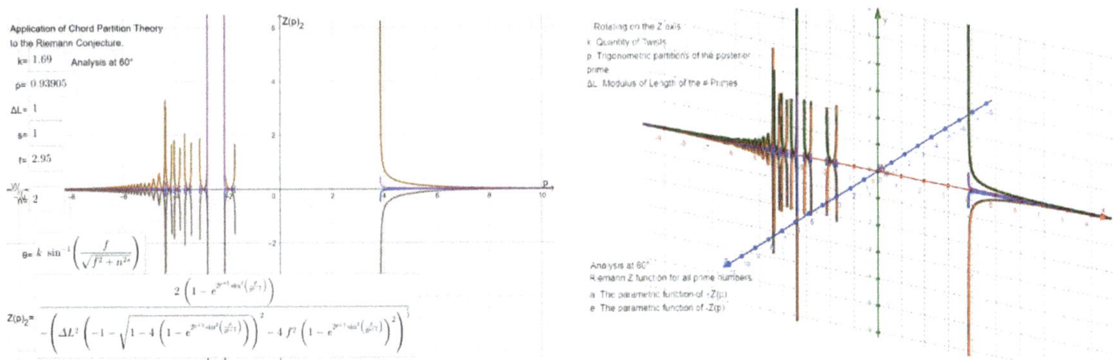

2D 3D

Figure 2.15: Riemann Z Function In The Plane Of Real Numbers.

The following graph is the equation that can be used with all values varying in time: $Z(p)_2 = \sum_{n=1}^{\infty}(1/n^s) = -[2*(1-e^{2^{(p+3)}*\sin^2(\theta/2^{p+1})})]$ $/\sqrt{[\Delta L^2*[-1-\sqrt{[1-4*(1-e^{2^{(p+3)}*\sin^2(\theta/2^{p+1})})]}]^2-4*f^2*(1-e^{2^{(p+3)}*\sin^2(\theta/2^{p+1})})]}$. With variable spiral angle; $\theta = 2{*}k{*}\arcsin\sqrt{[\Delta L/(2*n^s)]}$ or $\theta = -2{*}k{*}\arcsin\sqrt{[\Delta L/(2*n^s)]}$.

Which is a solution in the real number plane. In the 2D plane the equation generates dots and for it to be visible you have to select mark the trace in GeoGebra. In the 3D plane the equation generates arcs or circles depending on the values of **k** turns, and the initial spiral angle selected for your study of the Riemann Z Function. Where **p** are the trigonometric partitions, **n** is the posterior prime number,ΔL is the growth of the radius of the posterior prime number minus the previous one, **s** is the exponent of the real numbers, **f** is the height on the **Y** axis for the point starting point of the initial analysis with the theta spiral angle θ; in this case at 60°.

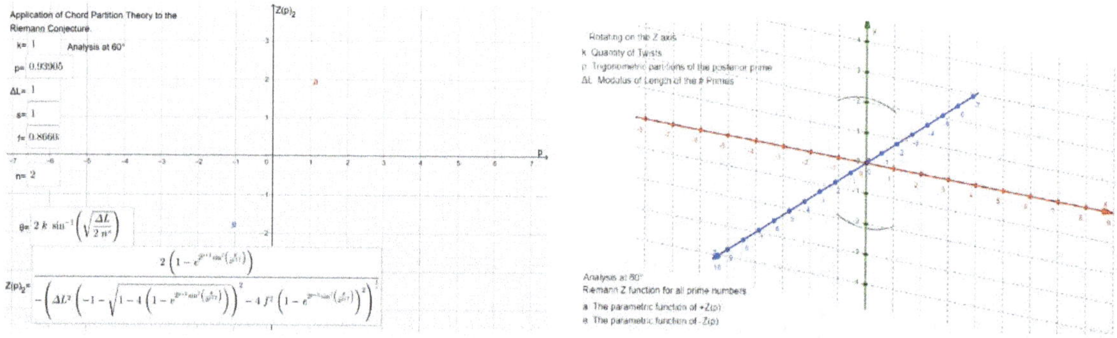

2D 3D

Figure 2.16: Riemann Z Function In The Plane Of Real Numbers.

The following graph is of the positive and negative equation $Z(p)_2$; with its **X** and **Y** components.

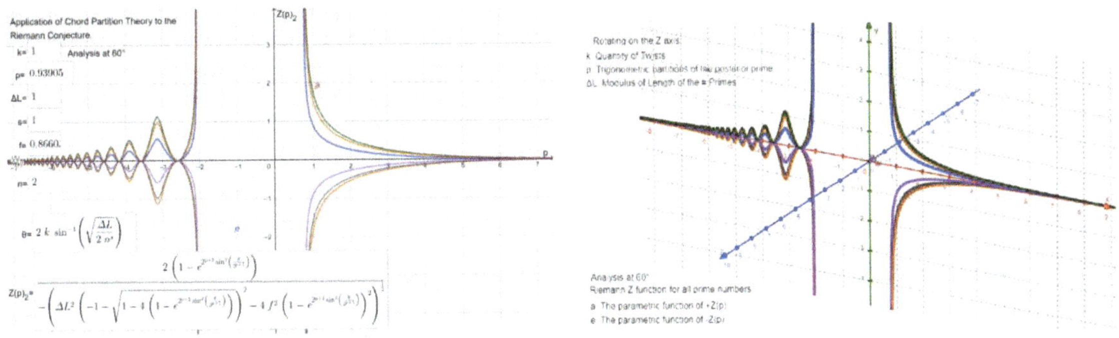

2D 3D

Figure 2.17: Riemann Z Function In The Plane Of Real Numbers.

The following graph is of the positive and negative equation $Z(p)_2$; with its **X** and **Y** components. With **k** turns, spiral angle θ and height **f**, varying in time. The rest of the constant variables.

2D 3D

Figure 2.18: Riemann Z Function In The Plane Of Real Numbers.

The following graph is the equation that can be used with all values varying in time:

$$Z(p)_3 = \sum_{n=1}^{\infty}(1/n^s) = \frac{[2*(1-e^{2(p+3)*\sin^2(\theta/2^{p+1})})]}{\sqrt{[\Delta L^2*[-1-\sqrt{[1-4*(1-e^{2(p+3)*\sin^2(\theta/2^{p+1})})]}]^2-4*f^2*(1-e^{2(p+3)*\sin^2(\theta/2^{p+1})})]}}.$$

With variable spiral angle;$\theta = k * \arcsin[f/\sqrt{(f^2 + n^{2s})}]$ Which is a solution in the real number plane. In the 2D plane the equation generates dots and for it to be visible you have to select mark the trace in GeoGebra. In the 3D plane the equation generates arcs or circles depending on the values of **k** turns, and the initial spiral angle selected for your study of the Riemann Z Function. Where **p** are the trigonometric partitions, **n** is the posterior prime number, ΔL is the growth of the radius of the posterior prime number minus the previous one, **s** is the exponent of the real numbers, **f** is the height on the **Y** axis for the point starting point of the initial analysis with the theta spiral angle θ; in this case at 60°. This graph is of the two positive and negative $Z(p)_3$ equations; with its **X** and **Y** components. Of the positive quadratic solution at 60° degrees.

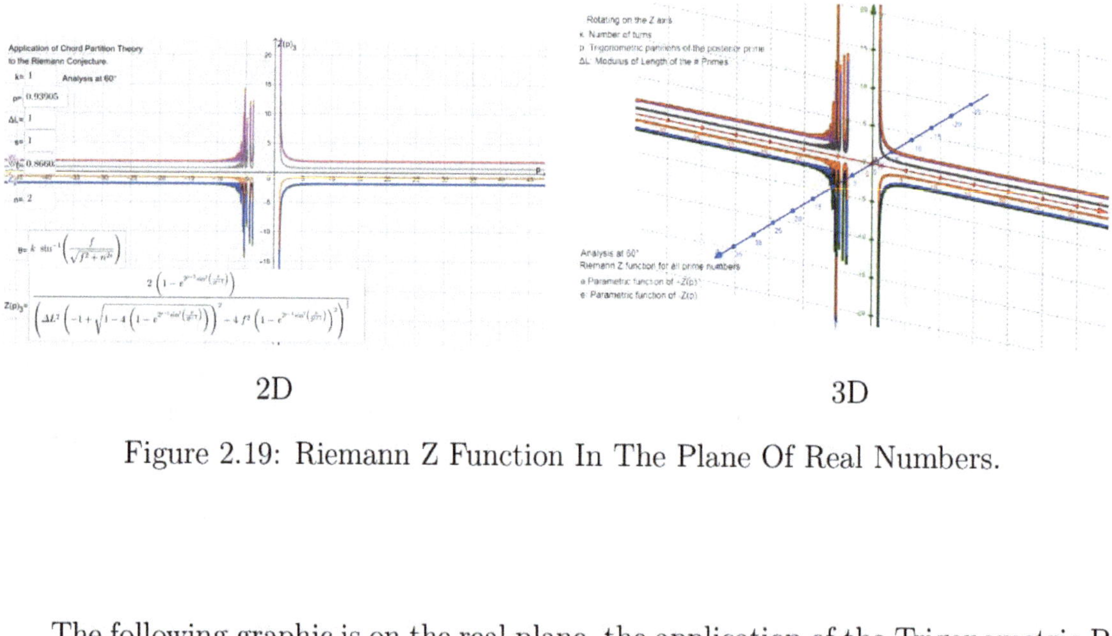

2D 3D

Figure 2.19: Riemann Z Function In The Plane Of Real Numbers.

The following graphic is on the real plane, the application of the Trigonometric Partitions of the Chord to the Riemann Z Function, of the solution of the quadratic equation, with the equation of the spiral angle φ.In this analysis, the equation obtained for the Riemann Hypothesis at 90° is used.

$Z(p) = \sum_{n=1}^{\infty} (1/n^s) = \sqrt{[e^{[2^{(p+3)} * \sin^2(\theta/2^{(p+1)})]} - 1]/\Delta L}.$

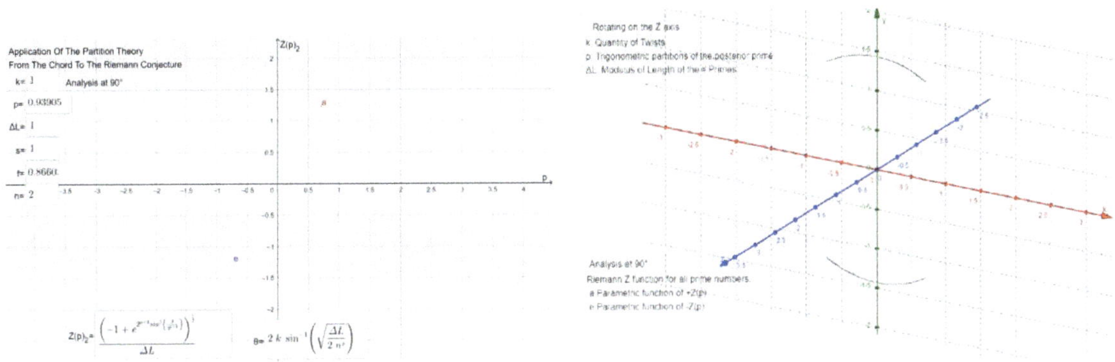

2D 3D

Figure 2.20: Riemann Z Function In The Plane Of Real Numbers.

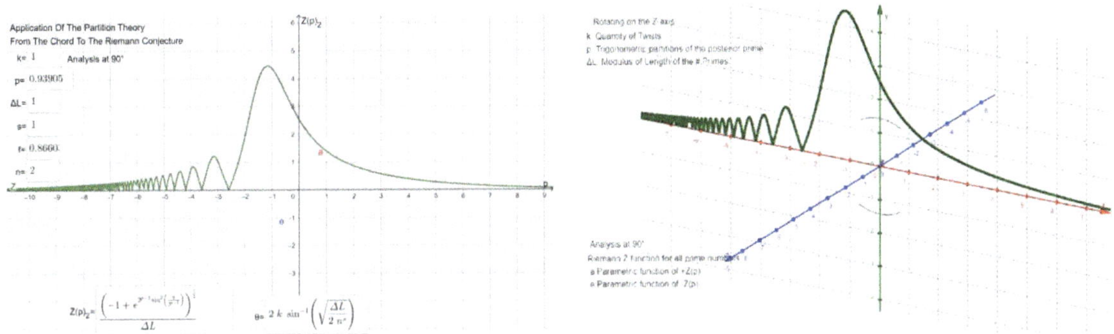

2D 3D

Figure 2.21: Riemann Z Function In The Plane Of Real Numbers.

In this section, With all the previous graphs, they can be transformed into this last type of graph by simply changing f=0, for the Cartesian plane of real numbers; for this to happen, it depends on the equation of the spiral angle being used. When changing the variable of a mathematical function from the Cartesian plane of real numbers to the complex number plane, its shape of graph also changes. Similarly, when we change the variable of a function in the same plane, the shape of graph also changes.

2.2 The Equation Of The Riemann Zeta Function In The Plane Of Complex Numbers.

The same four equations that have been developed in the plane of the real numbers will be graphed in the plane of the complex numbers where we will observe that all of them generate their own graphs in the plane of the complex numbers, being them a solution for Riemann´s Hypothesis or the Riemann Z Function. These four equations can generate other additional equations depending on the equations we use for the spiral angle. Rewriting them in a general way will obtain: Where **c** determines the size of the graph and **g** the constant where you want to place the graph on the x-axis. In this study **n** and ΔL must be different from zero.

Negative quadratic solution: The graph in the complex plane will be composed of two halves, one positive $+Z(\Delta L)_1$ and the other negative $-Z(\Delta L)_2$. Mathematical equation: $\varphi_0 = 120° = 2 * \pi/3$

$$Z(\Delta L)_1 = c * [2 * (1 - e^{2^{(p+3)} * \sin^2(\varphi/2^{p+1})})] / \sqrt{[\Delta L^2 * [-1 - \sqrt{[1 - 4 * 1 - e^{2^{(p+3)} * \sin^2(\varphi/2^{p+1})})]]^2 - 4 * f^2 * (1 - e^{2^{(p+3)} * \sin^2(\varphi/2^{p+1})})]} + g.$$

The negative square root solution: $Z(\Delta L)_2 = -c * [2 * (1 - e^{2^{(p+3)} * \sin^2(\varphi/2^{p+1})})]$

$$/ \sqrt{[\Delta L^2 * [-1 - \sqrt{[1 - 4 * (1 - e^{2^{(p+3)} * \sin^2(\varphi/2^{p+1})})]]^2 - 4 * f^2 * (1 - e^{2^{(p+3)} * \sin^2(\varphi/2^{p+1})})]} + g.$$

Positive quadratic solution: The graph in the complex plane will be composed of two halves, one positive $+Z(\Delta L)_3$ and the other negative $-Z(\Delta L)_4$.

$$Z(\Delta L)_3 = c * [2 * (1 - e^{2^{(p+3)} * \sin^2(\varphi/2^{p+1})})] / \sqrt{[\Delta L^2 * [-1 + \sqrt{[1 - 4 * (1 - e^{2^{(p+3)} * \sin^2(\varphi/2^{p+1})})]]^2 - 4 * f^2 * (1 - e^{2^{(p+3)} * \sin^2(\varphi/2^{p+1})})]} + g.$$

The negative square root solution:

$$Z(\Delta L)_4 = -c * [2 * (1 - e^{2^{(p+3)} * \sin^2(\varphi/2^{p+1})})] / \sqrt{[\Delta L^2 * [-1 + \sqrt{[1 - 4 * (1 - e^{2^{(p+3)} * \sin^2(\varphi/2^{p+1})})]]^2 - 4 * f^2 * (1 - e^{2^{(p+3)} * \sin^2(\varphi/2^{p+1})})]} + g.$$

A very important point is the equation of the spiral angle to be used in these equations since with each of them different behaviors are obtained, therefore, care must be taken when using the following equations; The first, which is the equation that numbers normally follow concerning their height **f**.Additionally, **f** is the variable that deforms the center of the graphs;It is the variable that is related to the Riemann sphere and its stereographic projection, with its Möbius map and its transformation.The spiral angle is: $\varphi = k * \arcsin[f/\sqrt{(f^2 + n^{2*s})}]$ or $\varphi = -k * \arcsin[f/\sqrt{(f^2 + n^{2*s})}]$, Likewise,the first, which is the equation that numbers normally follow concerning their height **f**, is: $\varphi = k * \arctan(f/n^s)$. The second, which are the angles that the prime numbers normally follow with their length module, is: $\varphi = 2 * k * \arcsin\sqrt{[\Delta L/(2 * n^s)]}$. or $\varphi = -2 * k * \arcsin\sqrt{[\Delta L/(2 * n^s)]}$.

The Riemann Z Function equations, of them can be rewritten depending on the variable that you want to obtain your graphs, however, some of them are dots,circular waves or invisible when creating them but their Z numerical values are the same. Examples of obtainable functions; Z(p), Z(n), $Z(\varphi)$, $Z(\zeta)$, Z(s), $Z(\Delta L)$.

In all these equations, the equivalent mathematical equations of the angle $\varphi = 2 * \pi * n = \omega * t = 2 * \pi * f * t = k * x - \omega * t$, where $k = 2 * \pi/\lambda$, can be used to perform analyses in terms of angular velocity and time, frequency and time, position and time, the number of waves k, and their wavelength λ. In this case, we will consider that the Z Function of the length module of the prime numbers is different to the Z function of the exponent s, Therefore, $Z(\Delta L) \neq Z(s)$, the numerical values of both equations can be observed in GeoGebra that the Z numerical values are equal with the difference that for $Z(\Delta L)$ the graph is visible;wich are circular waves and for Z(s) the graph is a dot or invisible. The behavior of an equation in the real number plane is different from the complex number plane. Examples if we substitute the spiral angle in the Riemann Z function, the equation would look like this: The negative quadratic solution, With:$\varphi = k * \arcsin[f/\sqrt{(f^2 + n^{2s})}]$ or $\varphi = k * \arctan(f/n^s)$.

$$Z(\Delta L)_1 = c * [2 * (1 - e^{2(p+3) * \sin^2(\varphi)/2^{p+1}})]$$
$$/\sqrt{[\Delta L^2 * [-1 - \sqrt{[1 - 4 * (1 - e^{2(p+3) * \sin^2(\varphi/2^{p+1})})]}]^2 - 4 * f^2 * (1 - e^{2(p+3) * \sin^2(\varphi/2^{p+1})})} + g.$$

This same angle will be used for the **X** and **Y** components.

The negative quadratic solution, With: $\zeta = k * \arctan(f/n^s)$.

$Z(\Delta L)_1 = c * [2 * (1 - e^{2(p+3)*\sin^2(\varsigma/2^{p+1})})]$
$/\sqrt{[\Delta L^2 * [-1 - \sqrt{[1 - 4 * (1 - e^{2(p+3)*\sin^2(\varsigma/2^{p+1})})]]^2 - 4 * f^2 * (1 - e^{2(p+3)*\sin^2(\varsigma/2^{p+1})})]} + g.$

This same angle will be used for the **X** and **Y** components. The following graphic is the application of the Trigonometric Partitions of the Chord to the Riemann Z Function, of the negative solution of the quadratic equation, with the equation of the spiral angle $\varphi = \zeta = k * \arcsin[f/\sqrt{(f^2 + n^{2s})}]$. In this analysis, the equation obtained for the Riemann Hypothesis at 45° is used. Located at the center of the Cartesian plane.

Let's observe something important: in order to obtain a graph in the complex number plane and provide a solution to the Riemann Zeta function, we must change variables from the real plane Z(p) to the complex number plane $Z(\Delta L)$. When we switch from the real Cartesian plane to the complex numbers, we need to change the Riemann Zeta function from trigonometric partitions p to the modulus of prime numbers ΔL.

In all these equations with f=0, which is equivalent to analyzing the Riemann Zeta function on the X-axis in summation form for Z(s), simpler equations are generated. With these, we can always create the graph of the Riemann Zeta function of various sizes and positions, either with real or complex numbers using $Z(\Delta L)$.

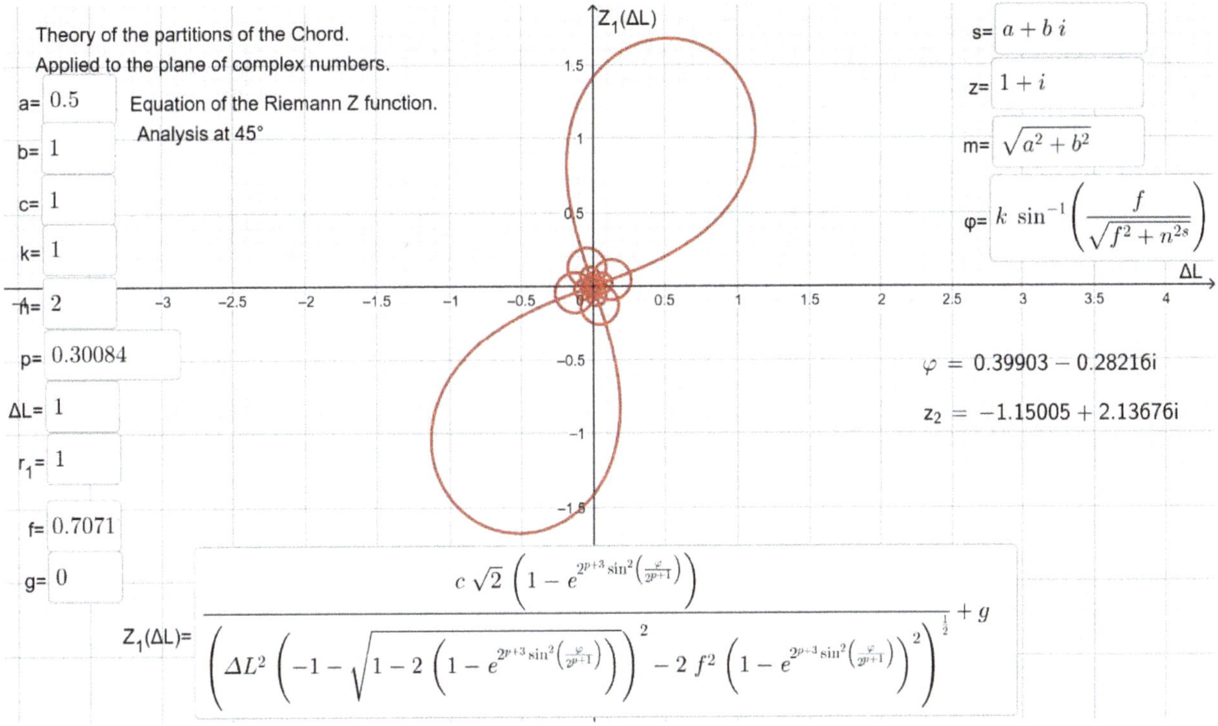

Figure 2.22: Riemann Z Function In The Plane Of Complex Numbers.

The following graphic is the application of the Trigonometric Partitions of the Chord to the Riemann Z Function, of the negative solution of the quadratic equation, with the equation of the spiral angle φ. In this analysis, the equation obtained for the Riemann Hypothesis at 45° is used. Moved on the **X** axis; with g=0.5 of the Cartesian plane. With these equations the non-trivial zeros in the plane of complex numbers can be obtained. If s=a+bi, are constants, **c** which determines the size of the graph which is changing in time from minus infinity to plus infinity, we can determine the non-trivial zeros, **k** turns, **p** trigonometric partitions, ΔL the length module of the prime numbers, **n** is the subsequent real prime number; are not varying in time, the equations $+Z(\Delta L)$ and $-Z(\Delta L)$ generate the complex numbers Z_2,Z_3,Z_4 and Z_5 in GeoGebra where we will get all non-trivial zeros with the two equations. Likewise, you can place all the variables changing in time and you will get more non-trivial zeros.

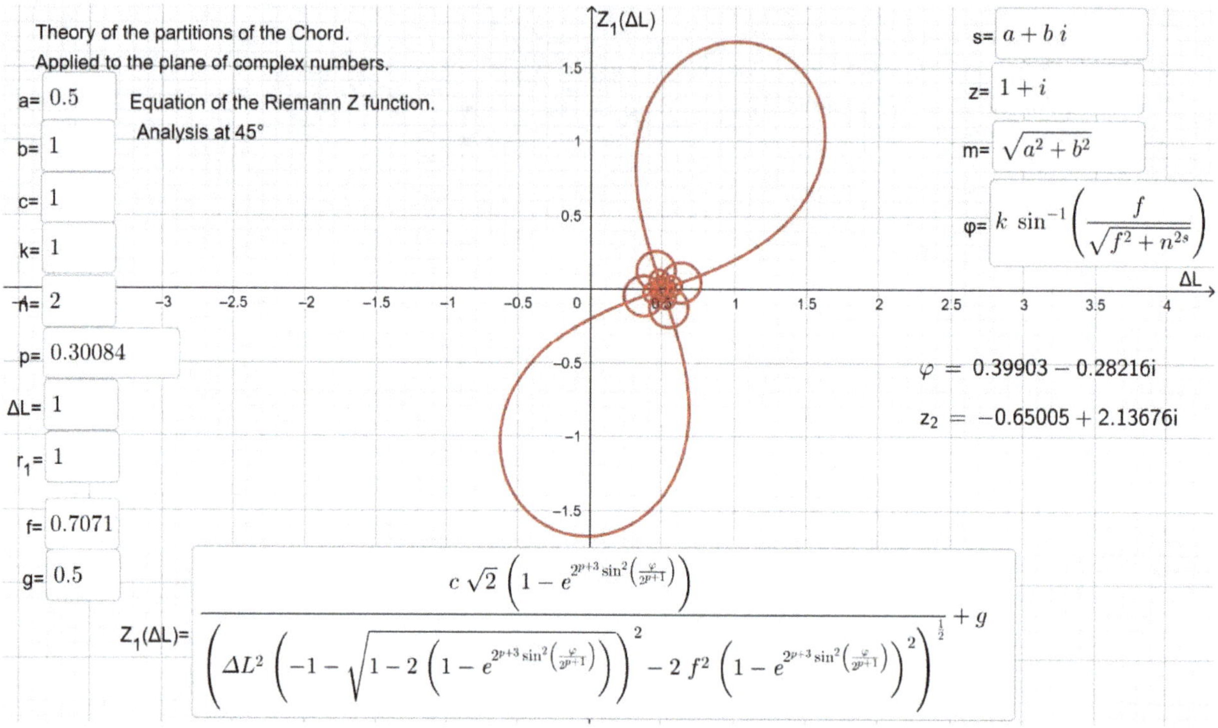

Figure 2.23: Riemann Z Function In The Plane Of Complex Numbers.

The following graphic is the application of the Trigonometric Partitions of the Chord to the Riemann Z Function, of the negative solution of the quadratic equation, with the equation of the spiral angle φ. In this analysis, the equation obtained for the Riemann Hypothesis at 45° is used. At the center of the Cartesian plane, with **b** and **f** varying in time.

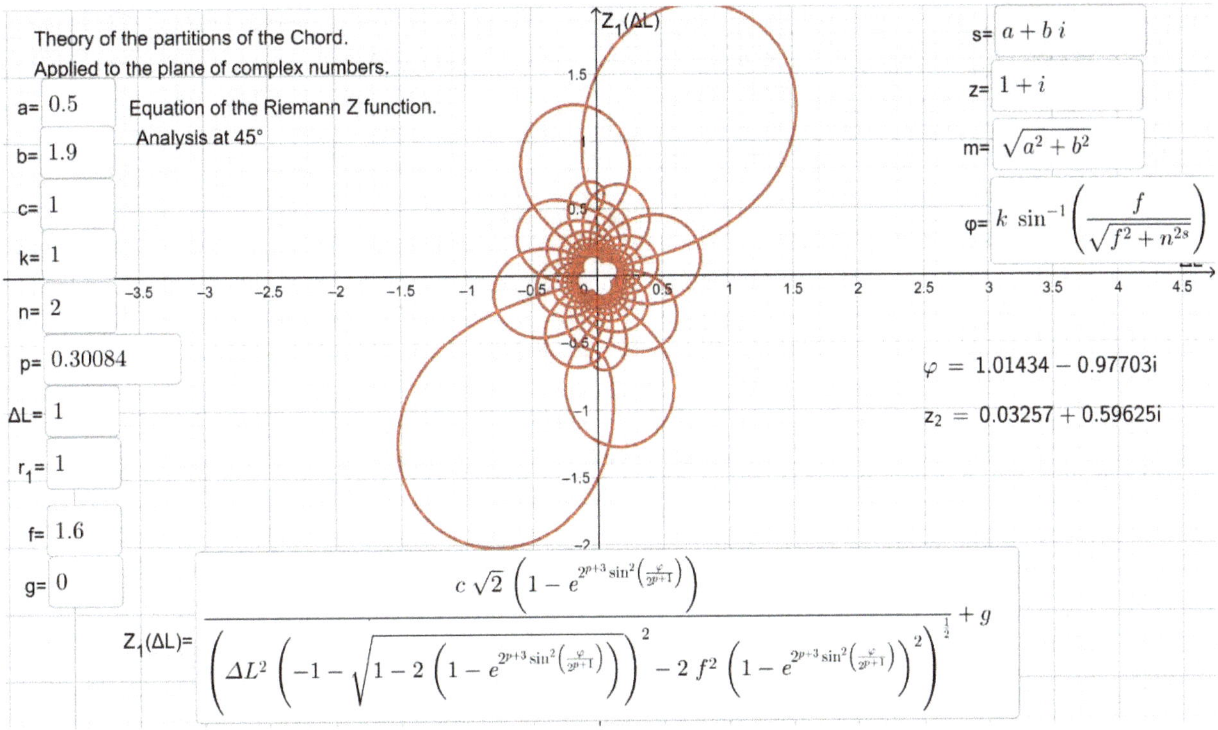

Figure 2.24: Riemann Z Function In The Plane Of Complex Numbers.

The following graphic is the application of the Trigonometric Partitions of the Chord to the Riemann Z Function, of the negative solution of the quadratic equation, with the equation of the spiral angle φ. In this analysis, the equation obtained for the Riemann Hypothesis at 45° is used. Moved on the **X** axis; with g=0.5 of the Cartesian plane, with **b** and **f** varying in time.

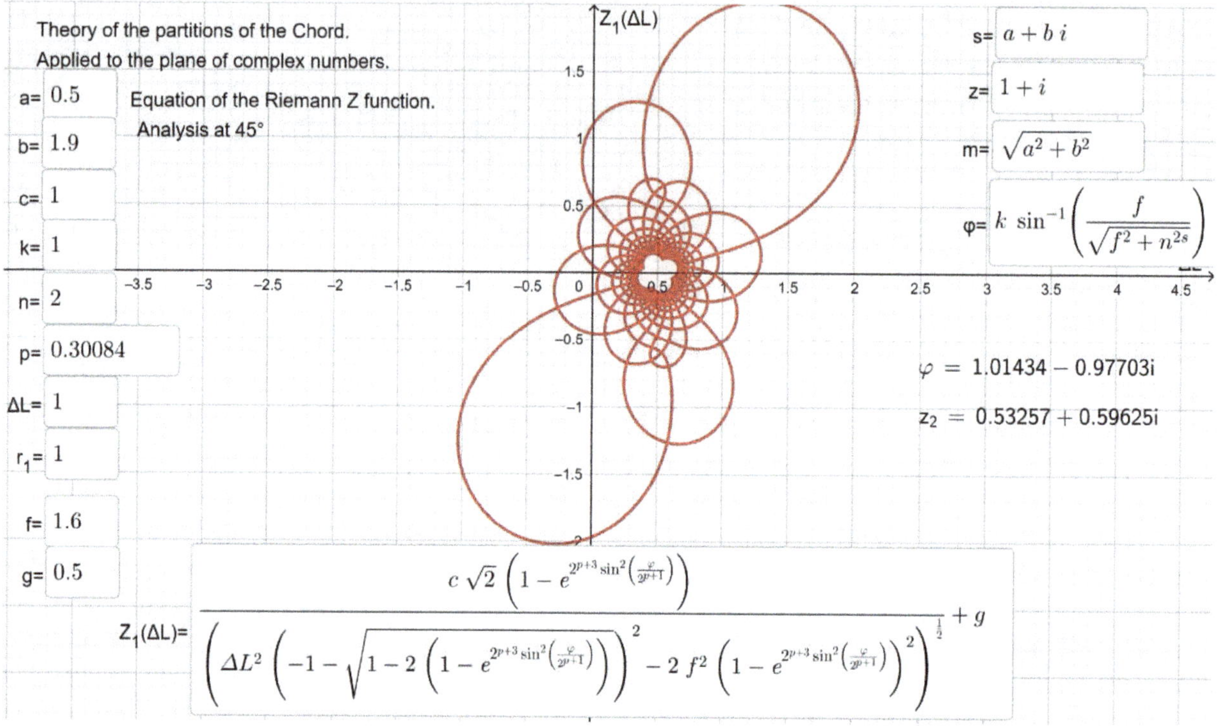

Figure 2.25: Riemann Z Function In The Plane Of Complex Numbers.

The following graphic is the application of the Trigonometric Partitions of the Chord to the Riemann Z Function, of the negative solution of the quadratic equation, with the equation of the spiral angle φ.In this analysis, the equation obtained for the Riemann Hypothesis at 45° is used. At the center of the Cartesian plane, with **b** and **f** varying in time.

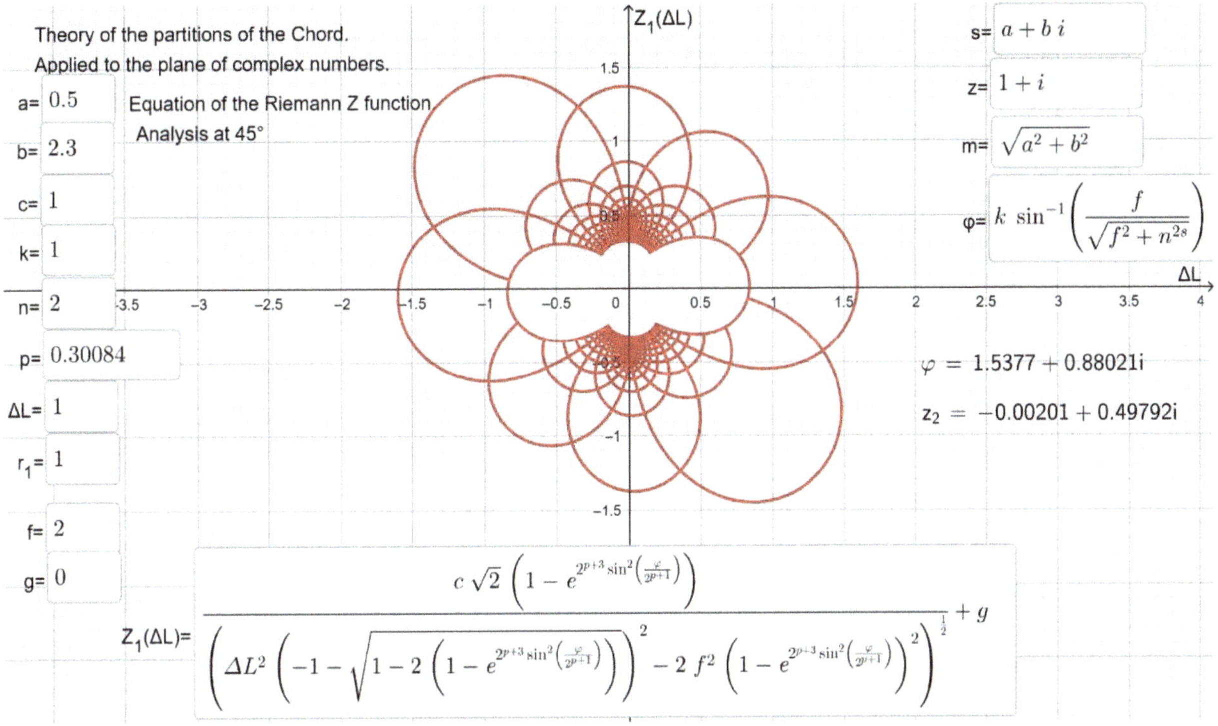

Figure 2.26: Riemann Z Function In The Plane Of Complex Numbers.

The following graphic is the application of the Trigonometric Partitions of the Chord to the Riemann Z Function, of the negative solution of the quadratic equation, with the equation of the spiral angle φ. In this analysis, the equation obtained for the Riemann Hypothesis at 45° is used. Moved on the **X** axis; with g=0.5 of the Cartesian plane, with **b** and **f** varying in time.

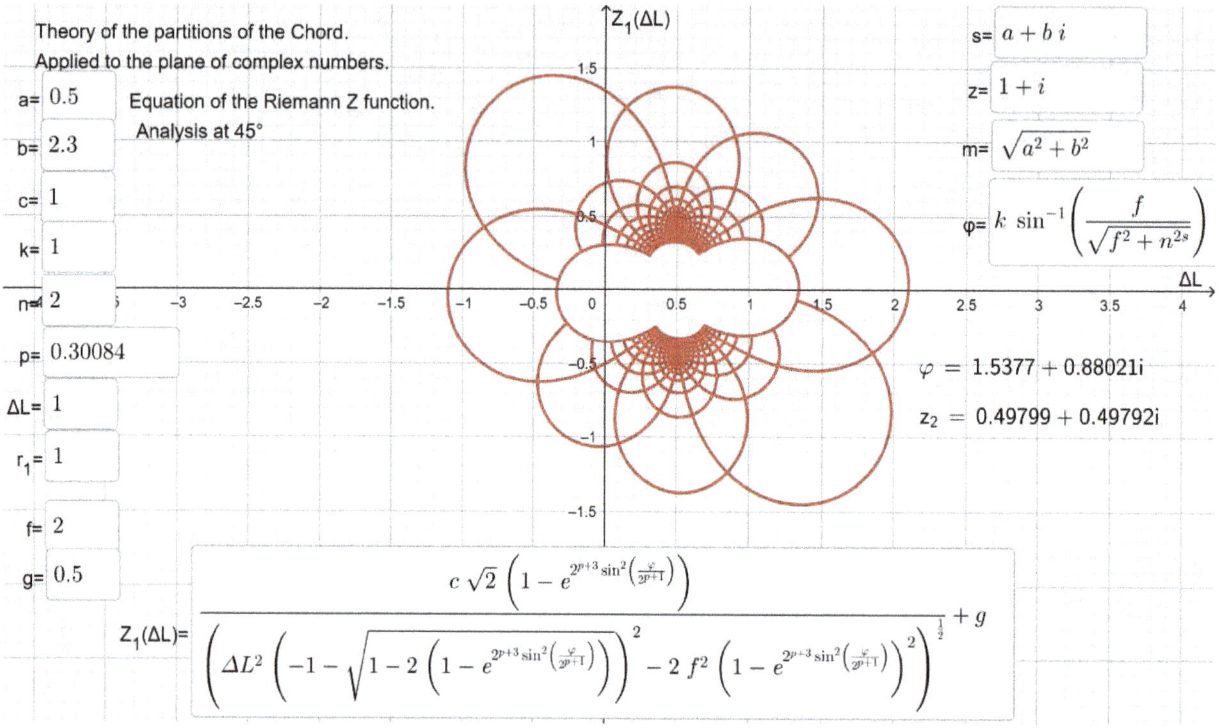

Figure 2.27: Riemann Z Function In The Plane Of Complex Numbers.

The following graphic is the application of the Trigonometric Partitions of the Chord to the Riemann Z Function, of the positive solution of the quadratic equation, with the equation of the spiral angle φ. In this analysis, the equation obtained for the Riemann Hypothesis at 45° is used. Located at the center of the Cartesian plane.

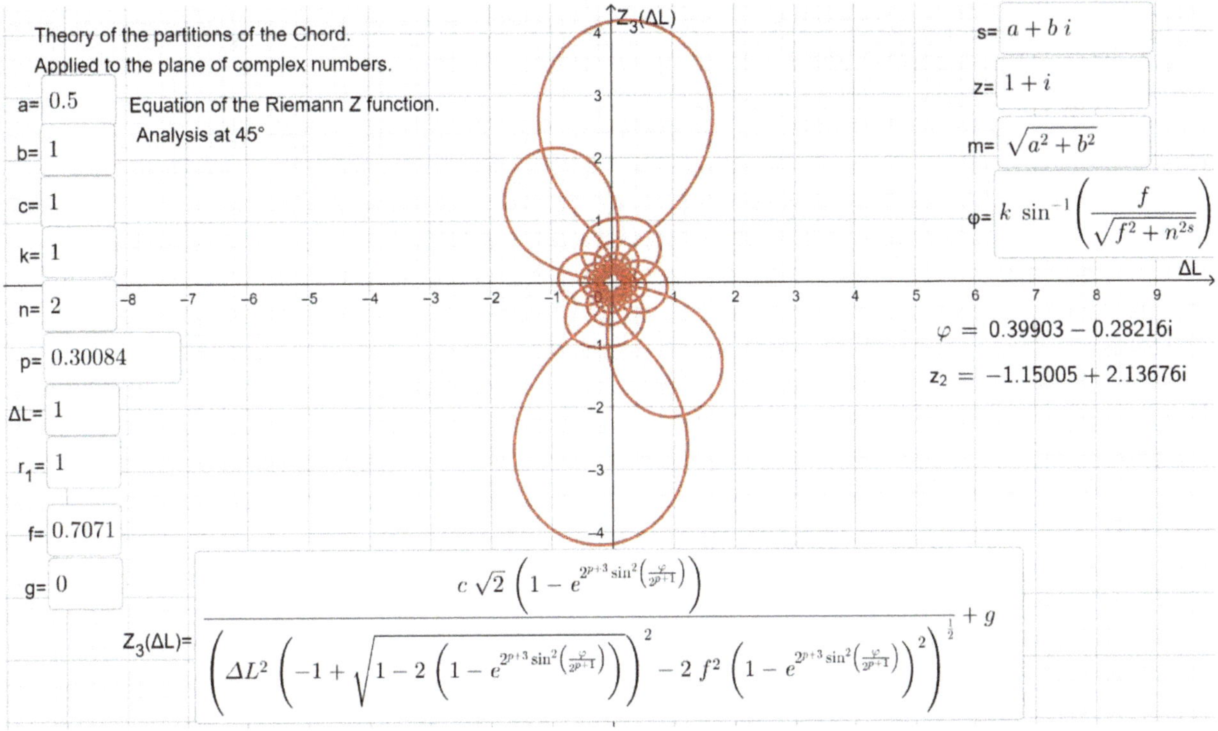

Figure 2.28: Riemann Z Function In The Plane Of Complex Numbers.

The following graphic is the application of the Trigonometric Partitions of the Chord to the Riemann Z Function, of the positive solution of the quadratic equation, with the equation of the spiral angle φ. In this analysis, the equation obtained for the Riemann Hypothesis at 45° is used. Moved on the **X** axis; with g=0.5 of the Cartesian plane. With these equations, the non-trivial zeros in the plane of complex numbers can be obtained. If s=a+bi, are constants, **c** which determines the size of the graph which is changing in time from minus infinity to plus infinity, we can determine the non-trivial zeros, **k** turns, **p** trigonometric partitions, ΔL the length module of the prime numbers, **n** is the subsequent real prime number; are not varying in time, the equations $+Z(\Delta L)$ and $-Z(\Delta L)$ generate the complex numbers Z_2,Z_3,Z_4 and Z_5 in GeoGebra where we will get all non-trivial zeros with the two equations. Likewise, you can place all the variables changing in time and you will get more non-trivial zeros.

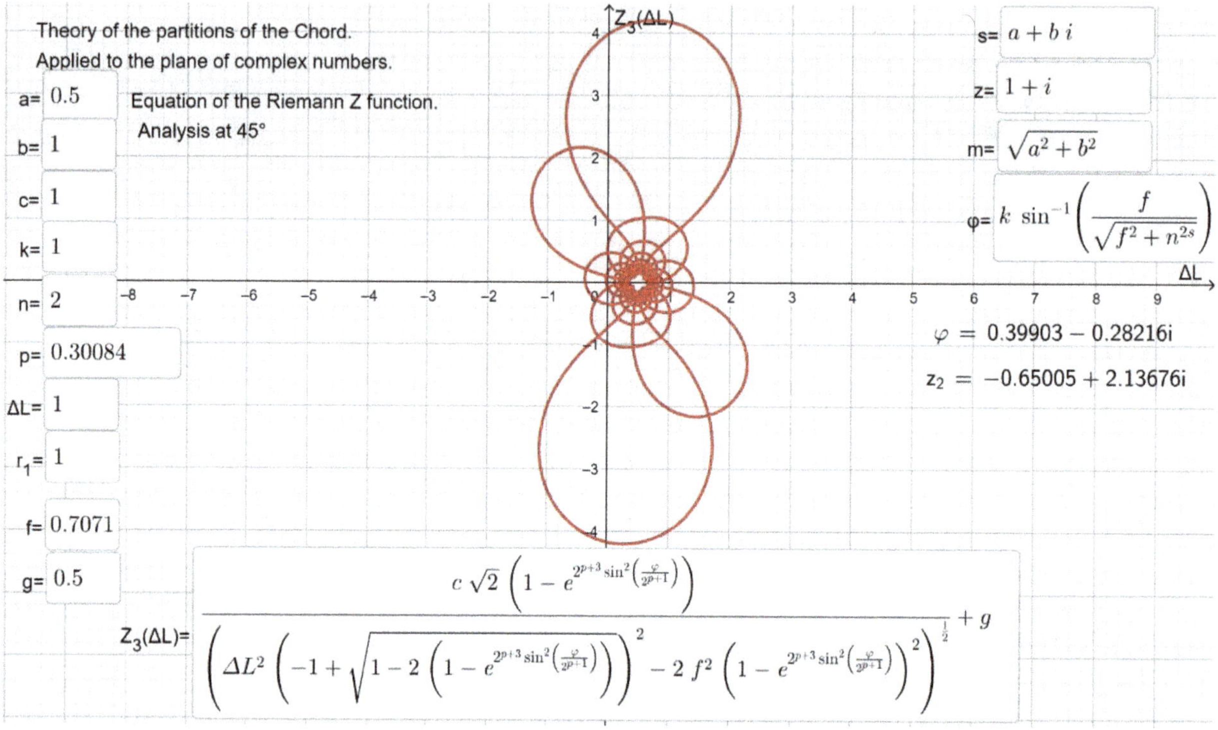

Figure 2.29: Riemann Z Function In The Plane Of Complex Numbers.

The following graphic is the application of the Trigonometric Partitions of the Chord to the Riemann Z Function, of the positive solution of the quadratic equation, with the equation of the spiral angle φ. In this analysis, the equation obtained for the Riemann Hypothesis at 45° is used. Located at the center of the Cartesian plane, with **b** and **f** varying in time.

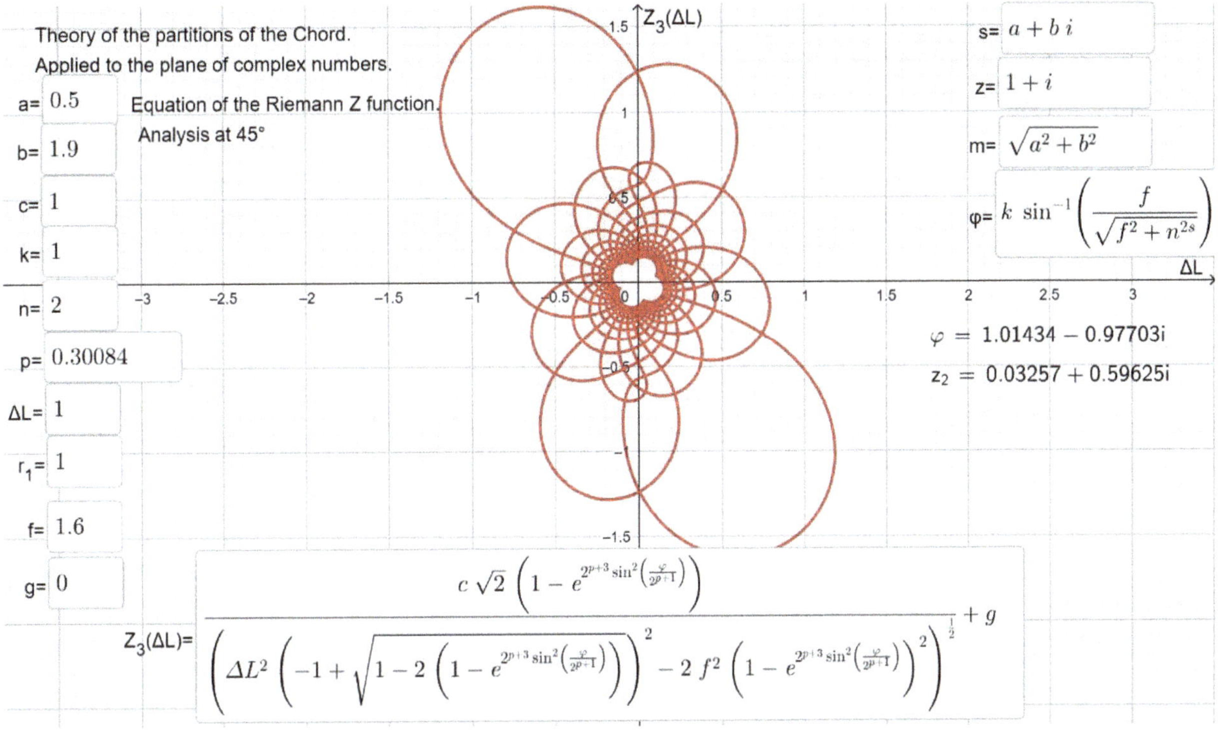

Figure 2.30: Riemann Z Function In The Plane Of Complex Numbers.

The following graphic is the application of the Trigonometric Partitions of the Chord to the Riemann Z Function, of the positive solution of the quadratic equation, with the equation of the spiral angle φ. In this analysis, the equation obtained for the Riemann Hypothesis at 45° is used. At the center of the Cartesian plane, with **b** and **f** varying in time.

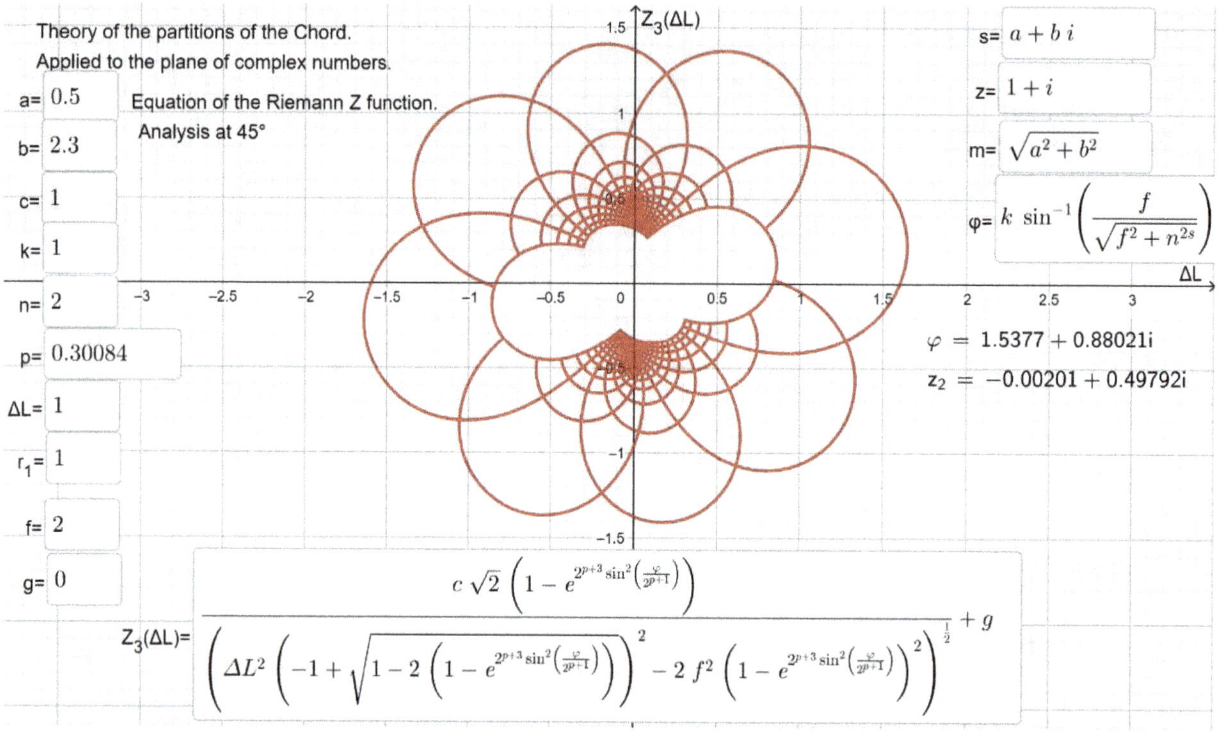

Figure 2.31: Riemann Z Function In The Plane Of Complex Numbers.

The following graphic is the application of the Trigonometric Partitions of the Chord to the Riemann Z Function, of the negative solution of the quadratic equation, with the equation of the spiral angle $\varphi = k * \arcsin[f/\sqrt{(f^2 + n^{2s})}]$. In this analysis, the equation obtained for the Riemann Hypothesis at 45° is used. Moved on the **X** axis; with g=0.5 of the Cartesian plane, with **b** and **f** varying in time.

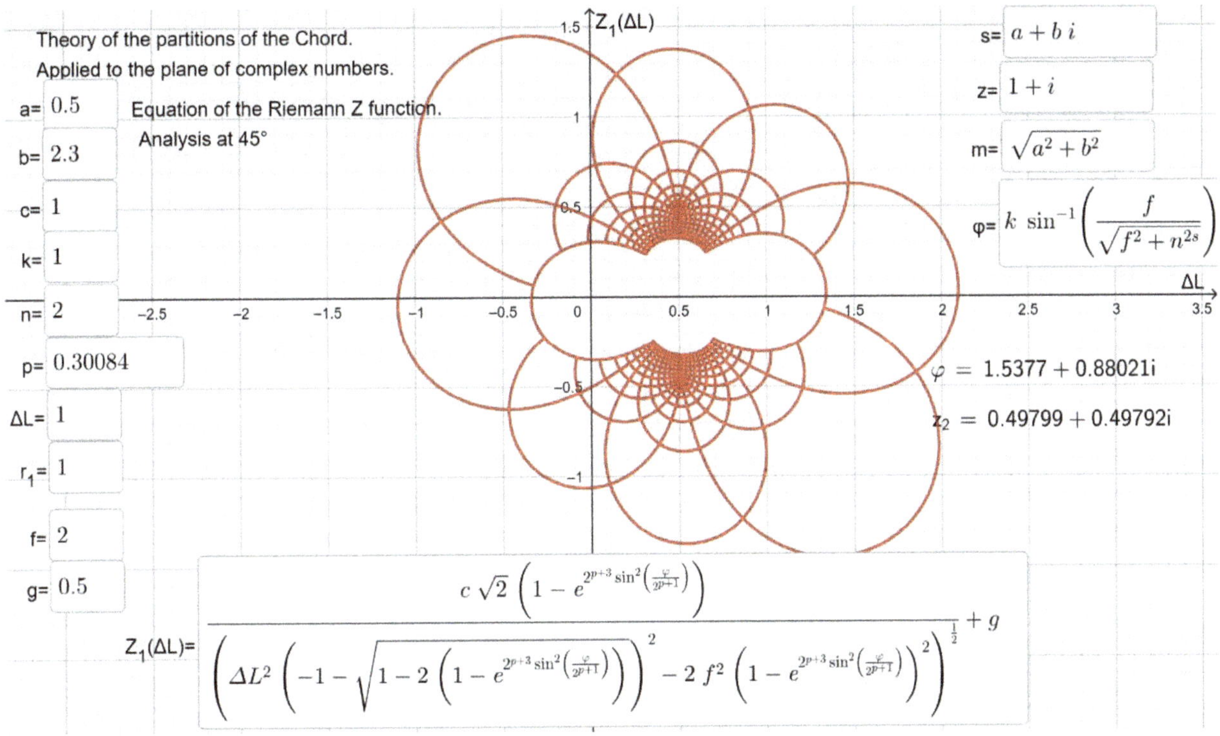

Figure 2.32: Riemann Z Function In The Plane Of Complex Numbers.

The following graphic is the application of the Trigonometric Partitions of the Chord to the Riemann Z Function, of the negative solution of the quadratic equation, with the equation of the spiral angle φ. In this analysis, the equation obtained for the Riemann Hypothesis at 60° is used. Located at the center of the Cartesian plane.

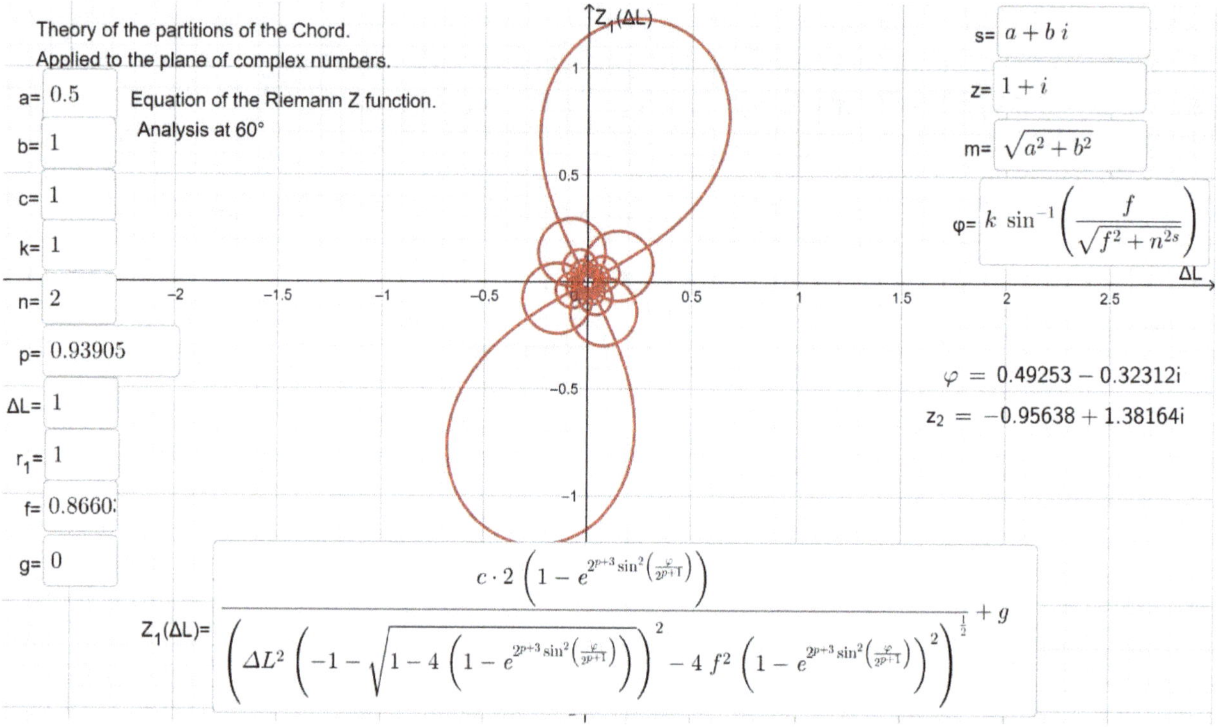

Figure 2.33: Riemann Z Function In The Plane Of Complex Numbers.

The following graphic is the application of the Trigonometric Partitions of the Chord to the Riemann Z Function, of the negative solution of the quadratic equation, with the equation of the spiral angle φ. In this analysis, the equation obtained for the Riemann Hypothesis at 60° is used. Moved on the **X** axis; with g=0.5 of the Cartesian plane. With these equations, the non-trivial zeros in the plane of complex numbers can be obtained. If s=a+bi, are constants, **c** which determines the size of the graph which is changing in time from minus infinity to plus infinity, we can determine the non-trivial zeros, **k** turns, **p** trigonometric partitions, ΔL the length module of the prime numbers, **n** is the subsequent real prime number; are not varying in time, the equations $+Z(\Delta L)$ and $-Z(\Delta L)$ generate the complex numbers Z_2,Z_3,Z_4 and Z_5 in GeoGebra where we will get all non-trivial zeros with the two equations. Likewise, you can place all the variables changing in time and you will get more non-trivial zeros.The following graph is the application of the Trigonometric Partitions of the String to the Riemann Z Function, of the negative solution of the quadratic equation, with the equation of the spiral angle φ.

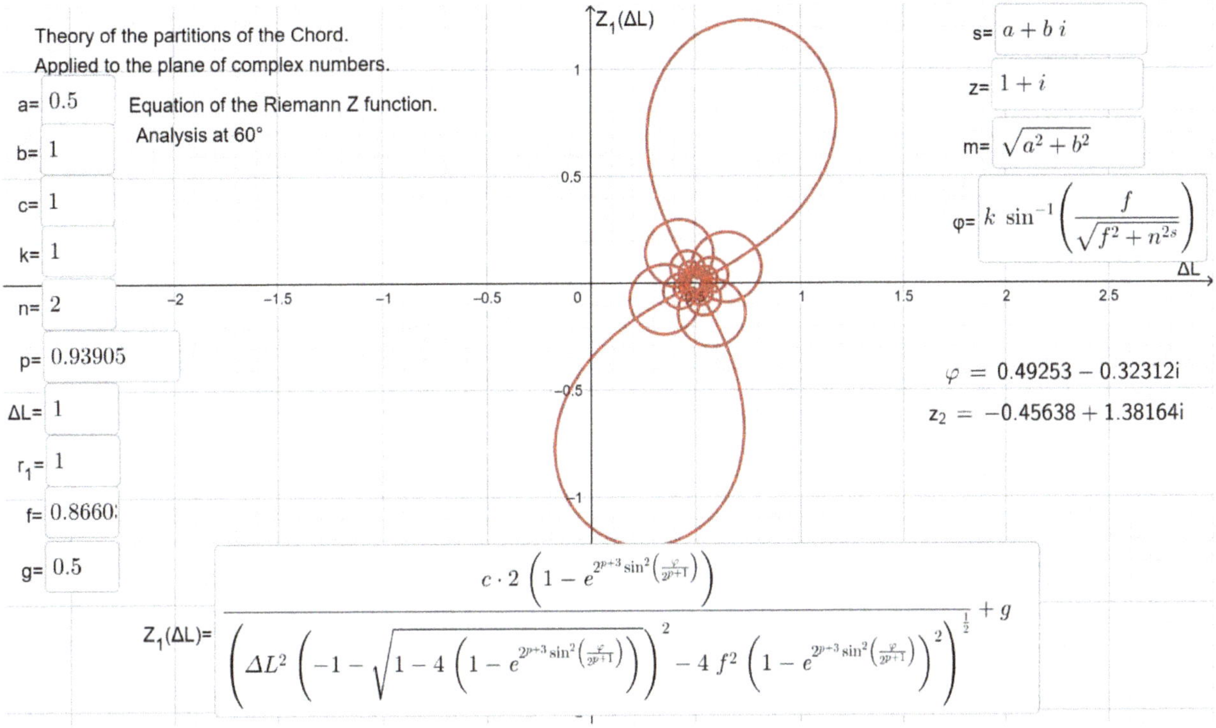

Figure 2.34: Riemann Z Function In The Plane Of Complex Numbers.

The following graphic is the application of the Trigonometric Partitions of the Chord to the Riemann Z Function, of the negative solution of the quadratic equation, with the equation of the spiral angle φ. In this analysis, the equation obtained for the Riemann Hypothesis at 60° is used. Moved on the **X** axis; with g=0.5 of the Cartesian plane, with **b** and **f** varying in time.

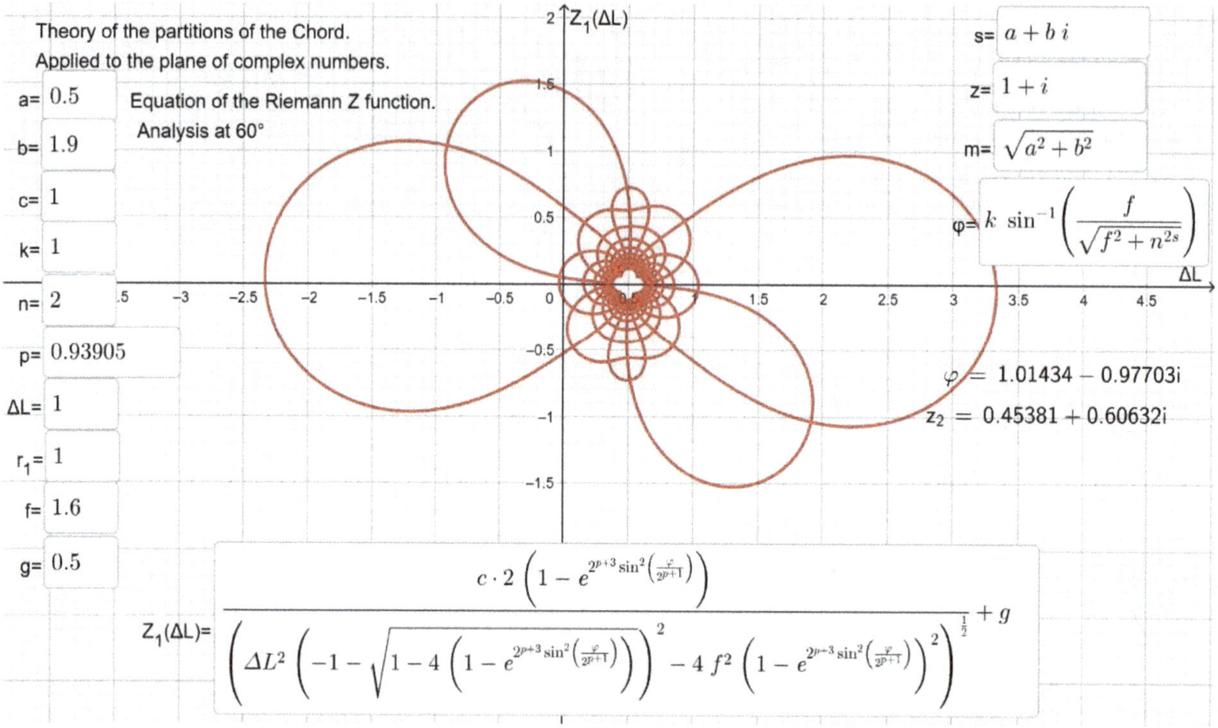

Figure 2.35: Riemann Z Function In The Plane Of Complex Numbers.

The following graphic is the application of the Trigonometric Partitions of the Chord to the Riemann Z Function, of the negative solution of the quadratic equation, with the equation of the spiral angle φ. In this analysis, the equation obtained for the Riemann Hypothesis at 60° is used. Located at the center of the Cartesian plane. With **b** and **f** varying in time.

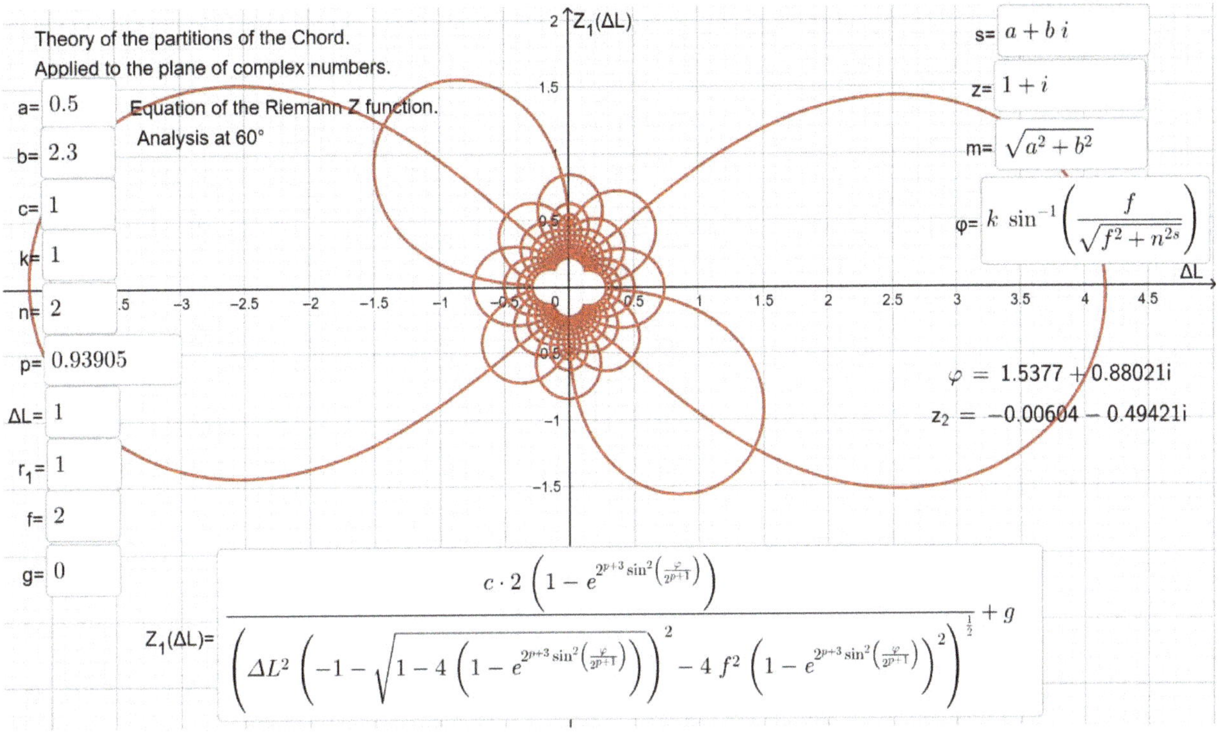

Figure 2.36: Riemann Z Function In The Plane Of Complex Numbers.

The following graphic is the application of the Trigonometric Partitions of the Chord to the Riemann Z Function, of the negative solution of the quadratic equation, with the equation of the spiral angle φ. In this analysis, the equation obtained for the Riemann Hypothesis at 60° is used. Moved on the **X** axis; with g=0.5 of the Cartesian plane, with **b** and **f** varying in time.

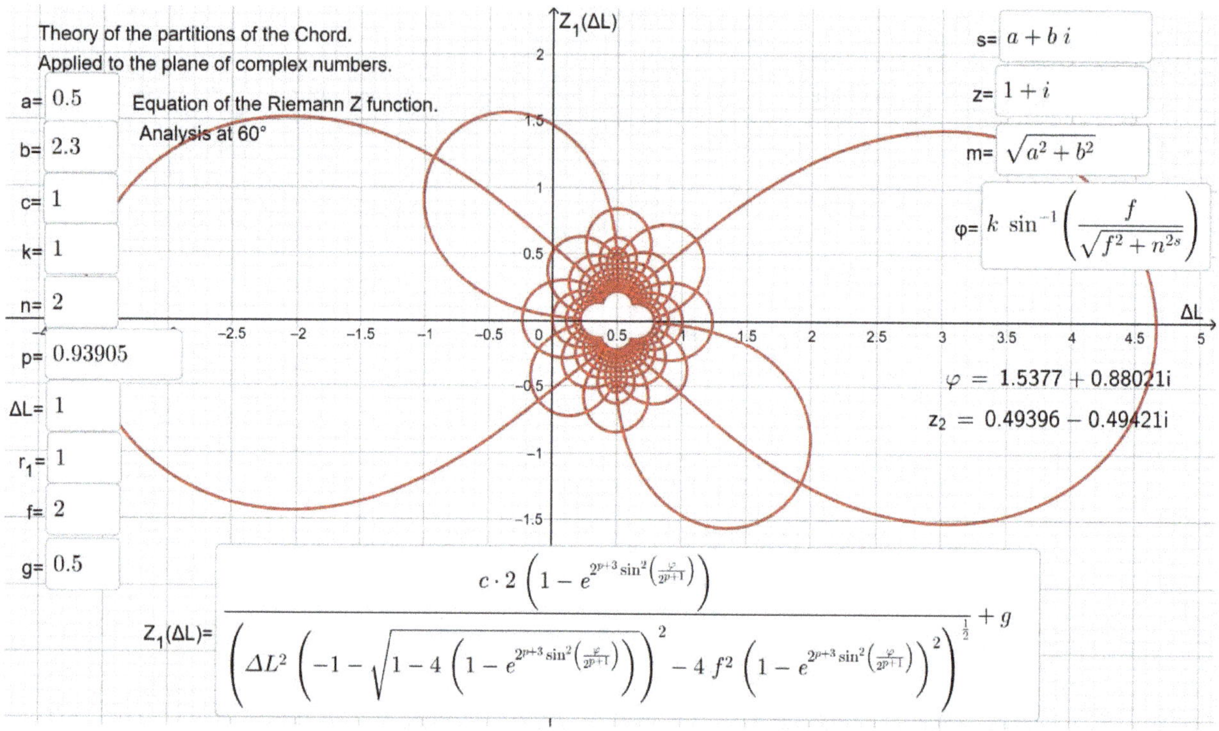

Figure 2.37: Riemann Z Function In The Plane Of Complex Numbers.

The following graphic is the application of the Trigonometric Partitions of the Chord to the Riemann Z Function, of the positive solution of the quadratic equation, with the equation of the spiral angle φ. In this analysis, the equation obtained for the Riemann Hypothesis at 60° is used. Located at the center of the Cartesian plane.

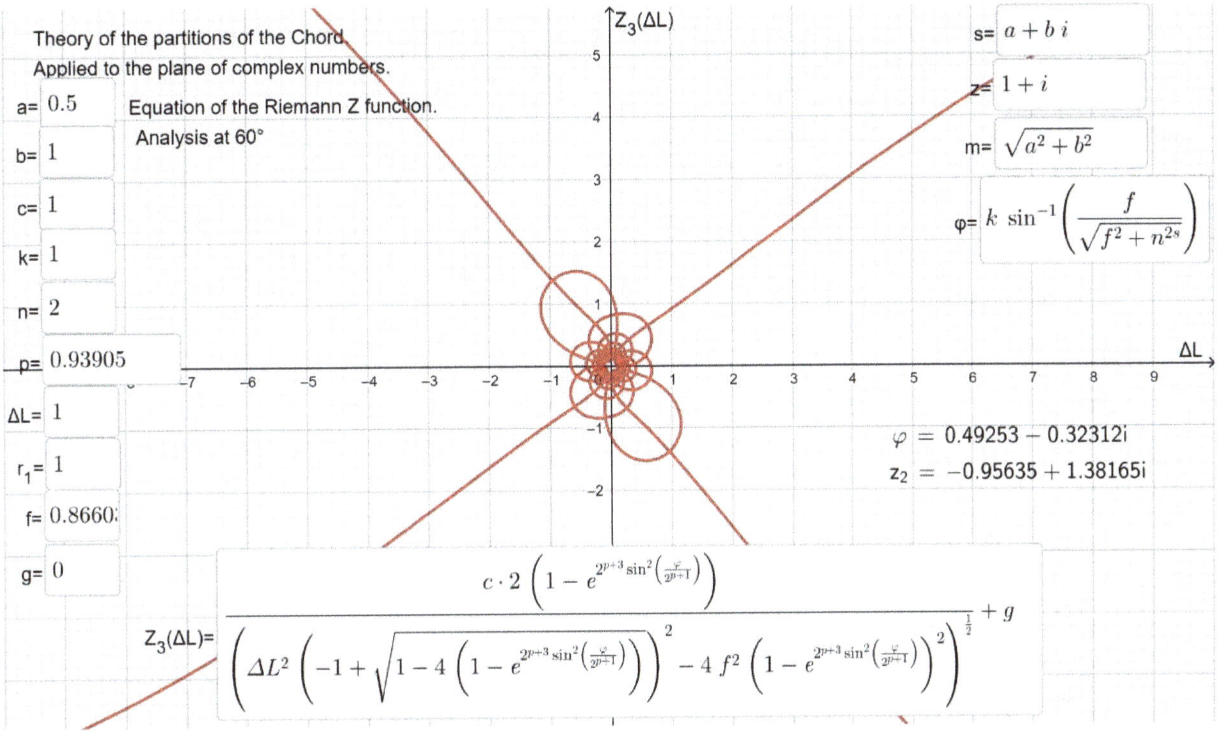

Figure 2.38: Riemann Z Function In The Plane Of Complex Numbers.

The following graphic is the application of the Trigonometric Partitions of the Chord to the Riemann Z Function, of the positive solution of the quadratic equation, with the equation of the spiral angle φ. In this analysis, the equation obtained for the Riemann Hypothesis at 60° is used. Moved on the X axis; with g=0.5 of the Cartesian plane.

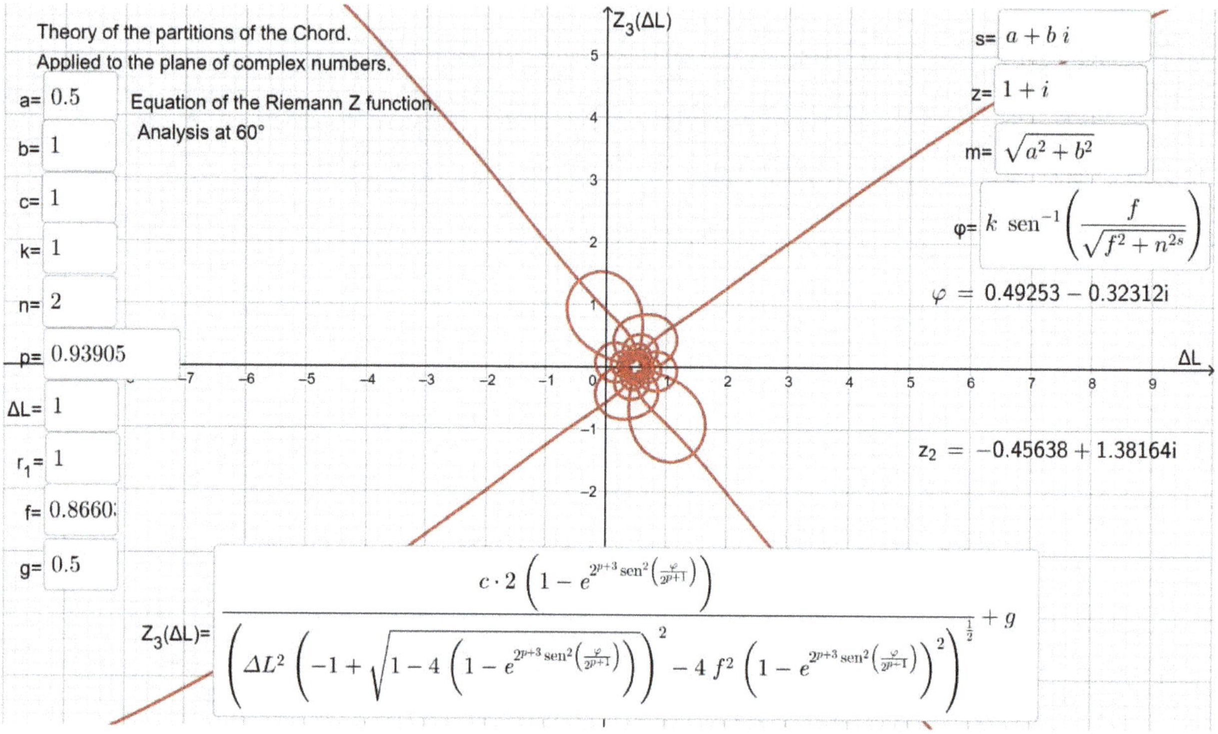

Figure 2.39: Riemann Z Function In The Plane Of Complex Numbers.

The following graphic is the application of the Trigonometric Partitions of the Chord to the Riemann Z Function, of the positive solution of the quadratic equation, with the equation of the spiral angle φ. In this analysis, the equation obtained for the Riemann Hypothesis at 60° is used. At the center of the Cartesian plane, with **b** and **f** varying in time.

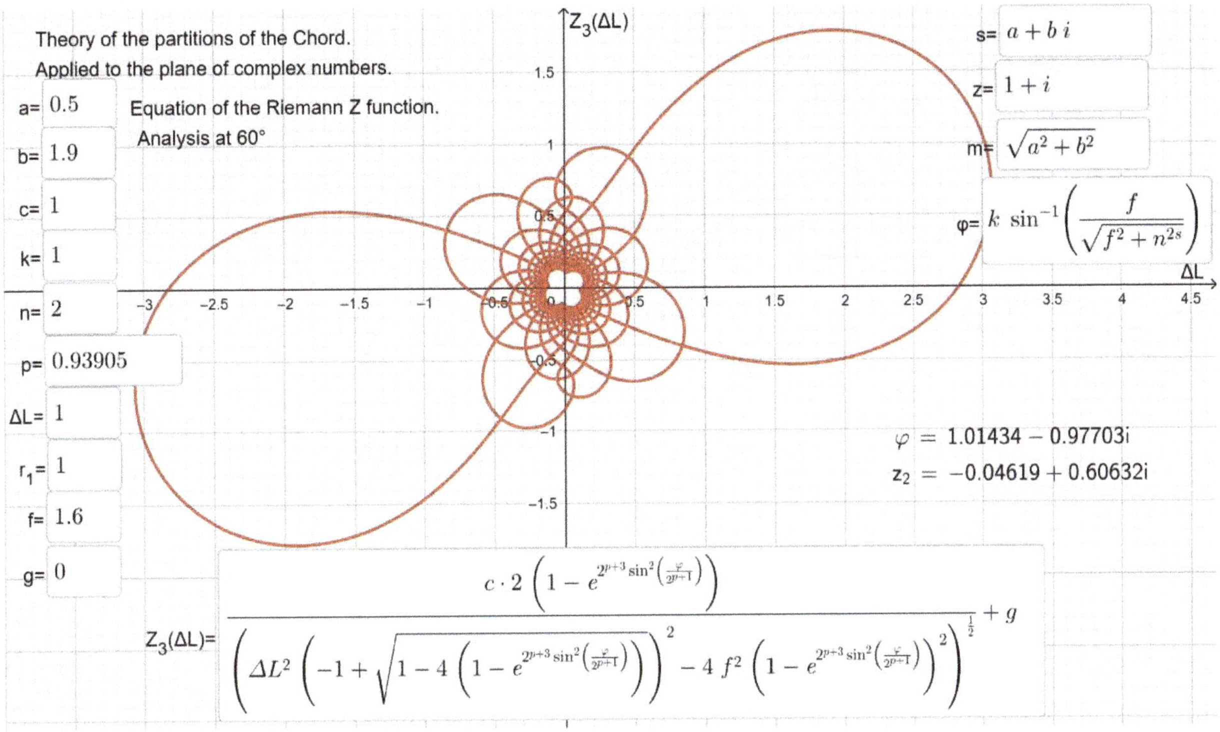

Figure 2.40: Riemann Z Function In The Plane Of Complex Numbers.

The following graphic is the application of the Trigonometric Partitions of the Chord to the Riemann Z Function, of the positive solution of the quadratic equation, with the equation of the spiral angle φ. In this analysis, the equation obtained for the Riemann Hypothesis at 60° is used. Moved on the **X** axis; with g=0.5 of the Cartesian plane, with **b** and **f** varying in time.

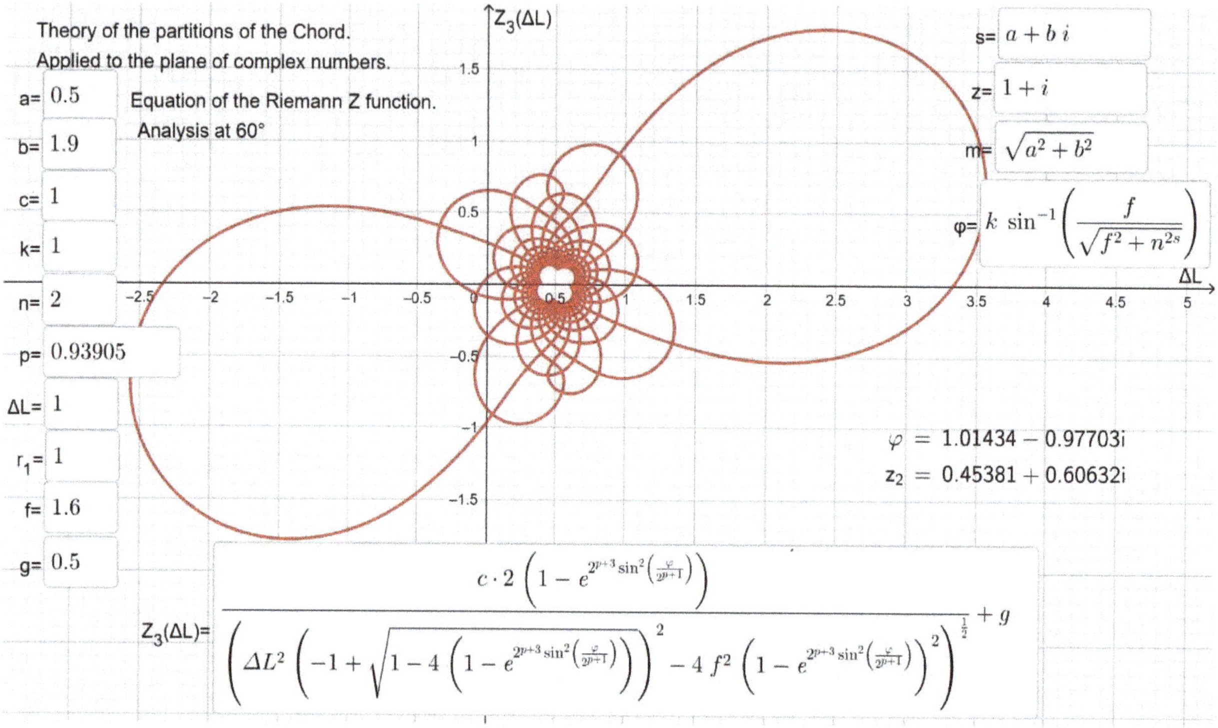

Figure 2.41: Riemann Z Function In The Plane Of Complex Numbers.

The following graphic is the application of the Trigonometric Partitions of the Chord to the Riemann Z Function, of the positive solution of the quadratic equation, with the equation of the spiral angle φ. In this analysis, the equation obtained for the Riemann Hypothesis at 60° is used. At the center of the Cartesian plane, with **b** and **f** varying in time.

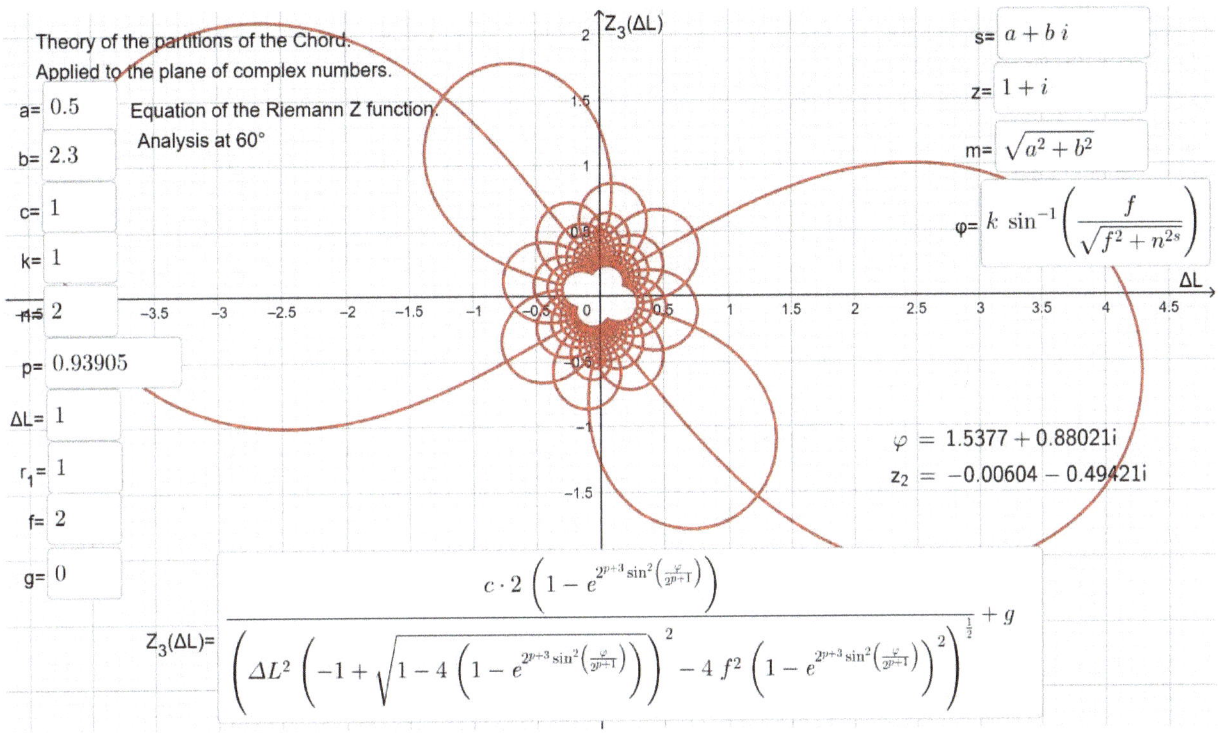

Figure 2.42: Riemann Z Function In The Plane Of Complex Numbers.

The following graphic is the application of the Trigonometric Partitions of the Chord to the Riemann Z Function, of the positive solution of the quadratic equation, with the equation of the spiral angle φ. In this analysis, the equation obtained for the Riemann Hypothesis at 60° is used. Moved on the X axis; with g=0.5 of the Cartesian plane, with **b** and **f** varying in time. With these equations, the non-trivial zeros in the plane of complex numbers can be obtained. If s=a+bi, are constants, **c** which determines the size of the graph which is changing in time from minus infinity to plus infinity, we can determine the non-trivial zeros, **k** turns, **p** trigonometric partitions, ΔL the length module of the prime numbers, **n** is the subsequent real prime number; are not varying in time, the equations $+Z(\Delta L)$ and $-Z(\Delta L)$ generate the complex numbers Z_3 and Z_4 in GeoGebra where we will get all non-trivial zeros with the two equations. Likewise, you can place all the variables changing in time and you will get more non-trivial zeros.

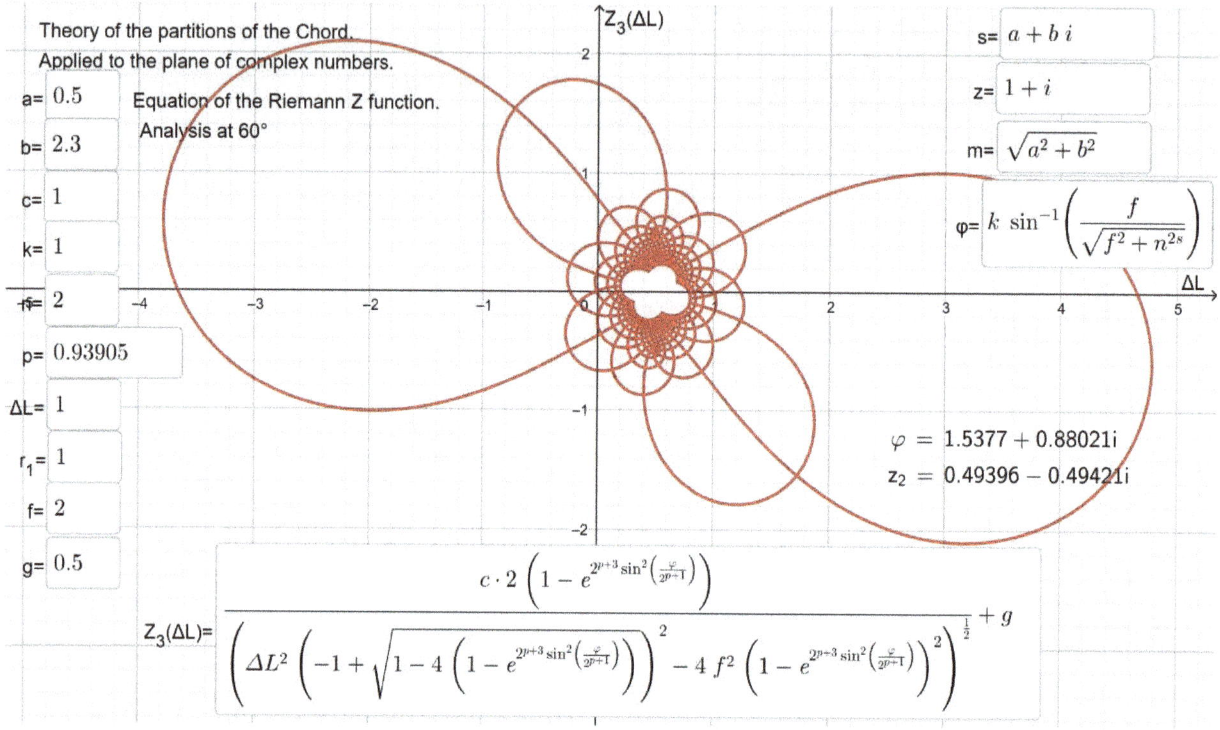

Figure 2.43: Riemann Z Function In The Plane Of Complex Numbers.

The following graphic is the application of the Trigonometric Partitions of the Chord to the Riemann Z Function, of the negative solution of the quadratic equation, with the equation of the spiral angle; $\varphi = 2 * k * \arcsin \sqrt{[\Delta L/(2 * n^s)]}$. In this analysis, the equation obtained for the Riemann Hypothesis at 60° is used. Located at the center of the Cartesian plane.

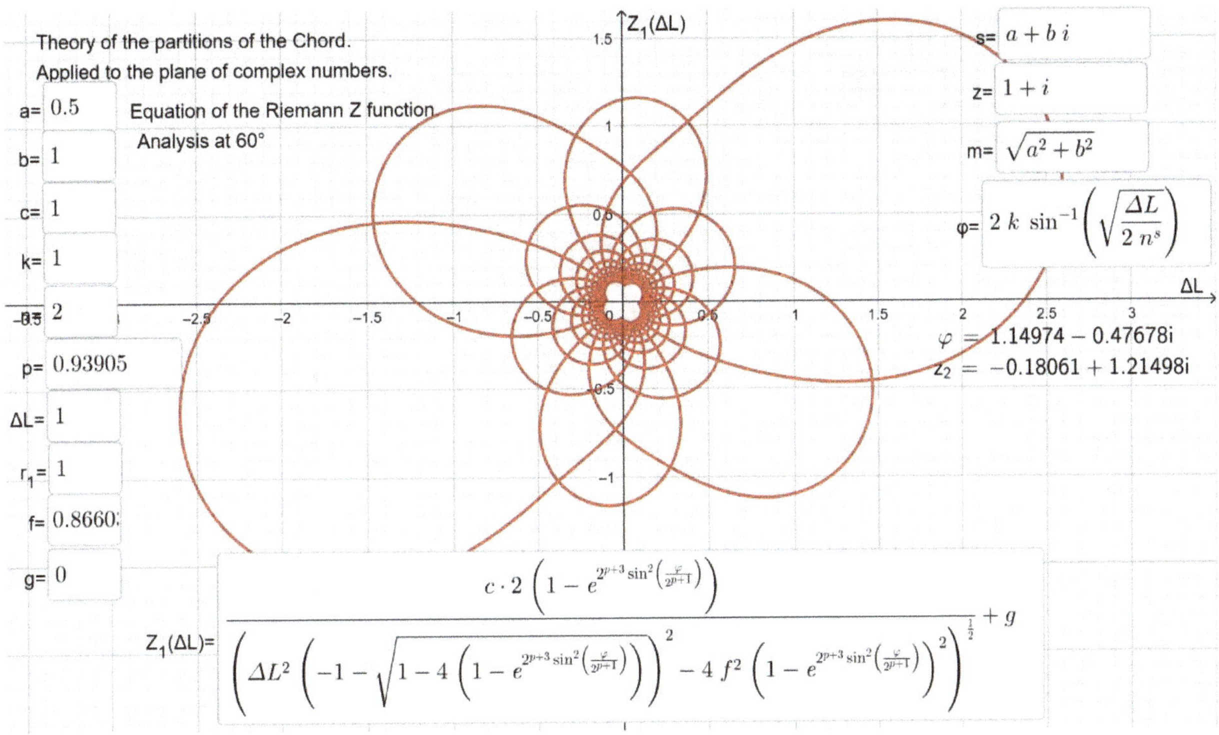

Figure 2.44: Riemann Z Function In The Plane Of Complex Numbers.

The following graphic is the application of the Trigonometric Partitions of the Chord to the Riemann Z Function, of the negative solution of the quadratic equation, with the equation of the spiral angle φ. In this analysis, the equation obtained for the Riemann Hypothesis at 60° is used. Moved on the X axis; with g=0.5 of the Cartesian plane.

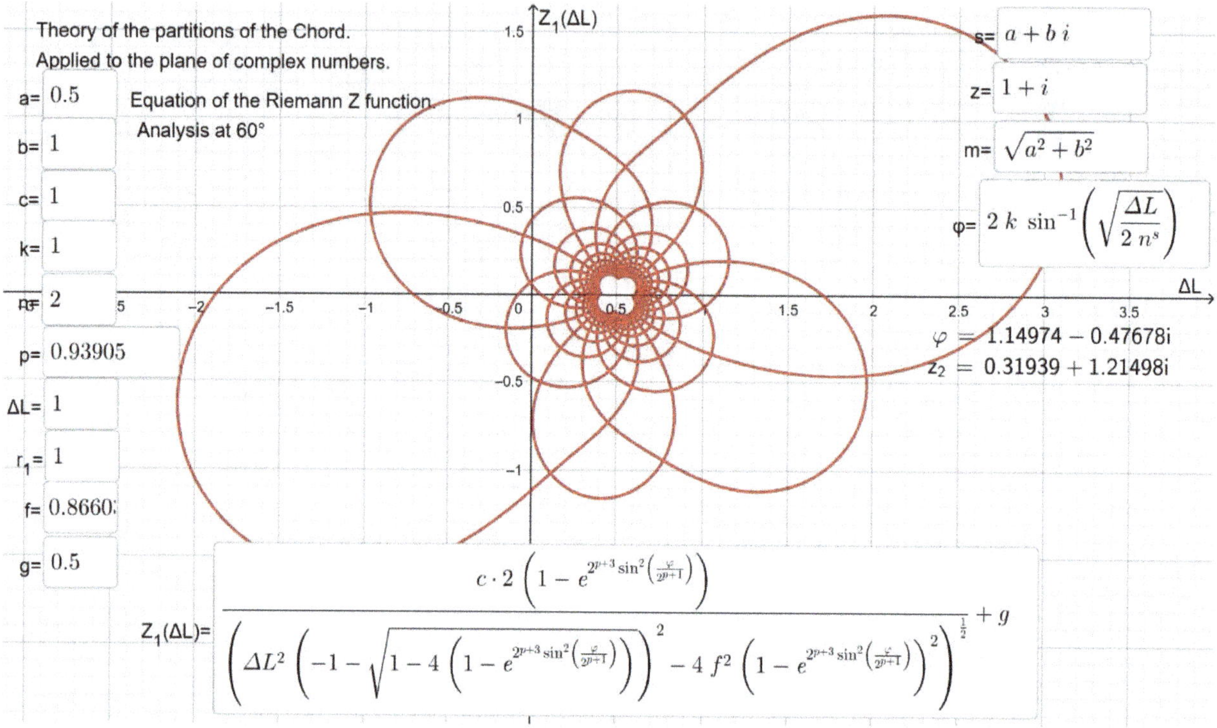

Figure 2.45: Riemann Z Function In The Plane Of Complex Numbers.

The following graphic is the application of the Trigonometric Partitions of the Chord to the Riemann Z Function, of the negative solution of the quadratic equation, with the equation of the spiral angle φ. In this analysis, the equation obtained for the Riemann Hypothesis at 60° is used. At the center of the Cartesian plane, with **b** and **f** varying in time.

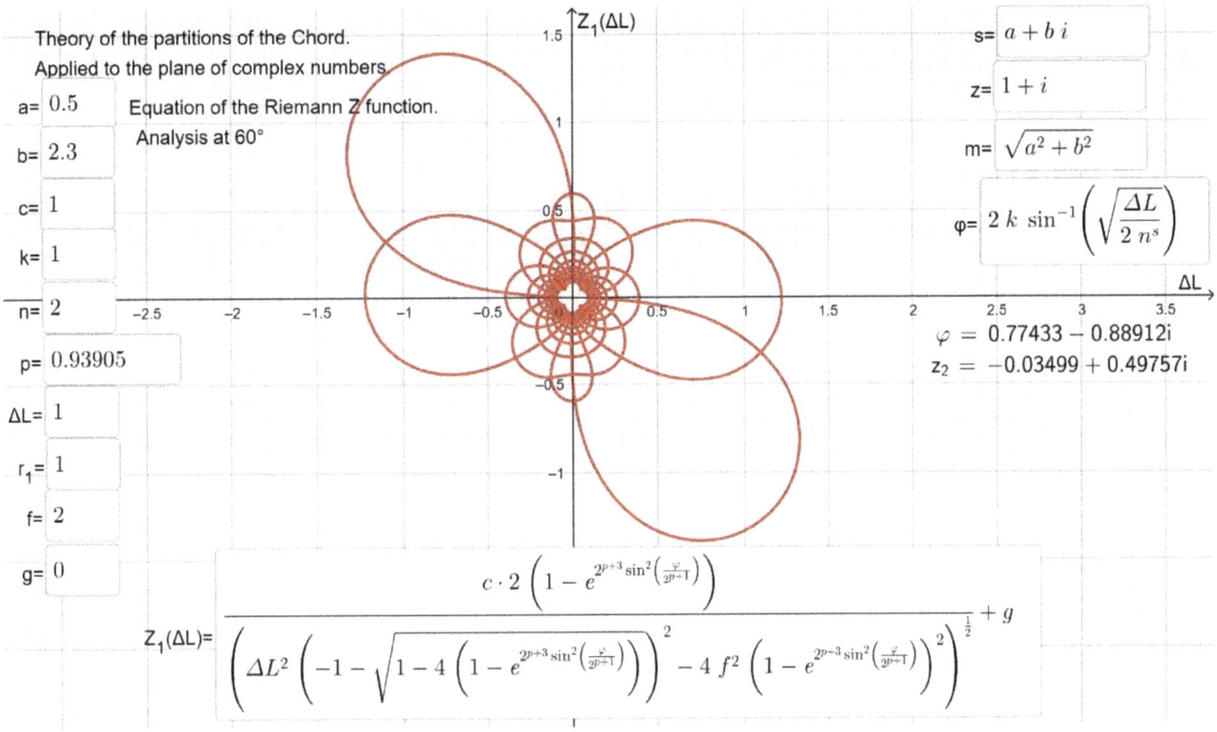

Figure 2.46: Riemann Z Function In The Plane Of Complex Numbers.

The following graphic is the application of the Trigonometric Partitions of the Chord to the Riemann Z Function, of the negative solution of the quadratic equation, with the equation of the spiral angle φ. In this analysis, the equation obtained for the Riemann Hypothesis at 60° is used. Moved on the **X** axis; with g=0.5 of the Cartesian plane, with **b** and **f** varying in time.

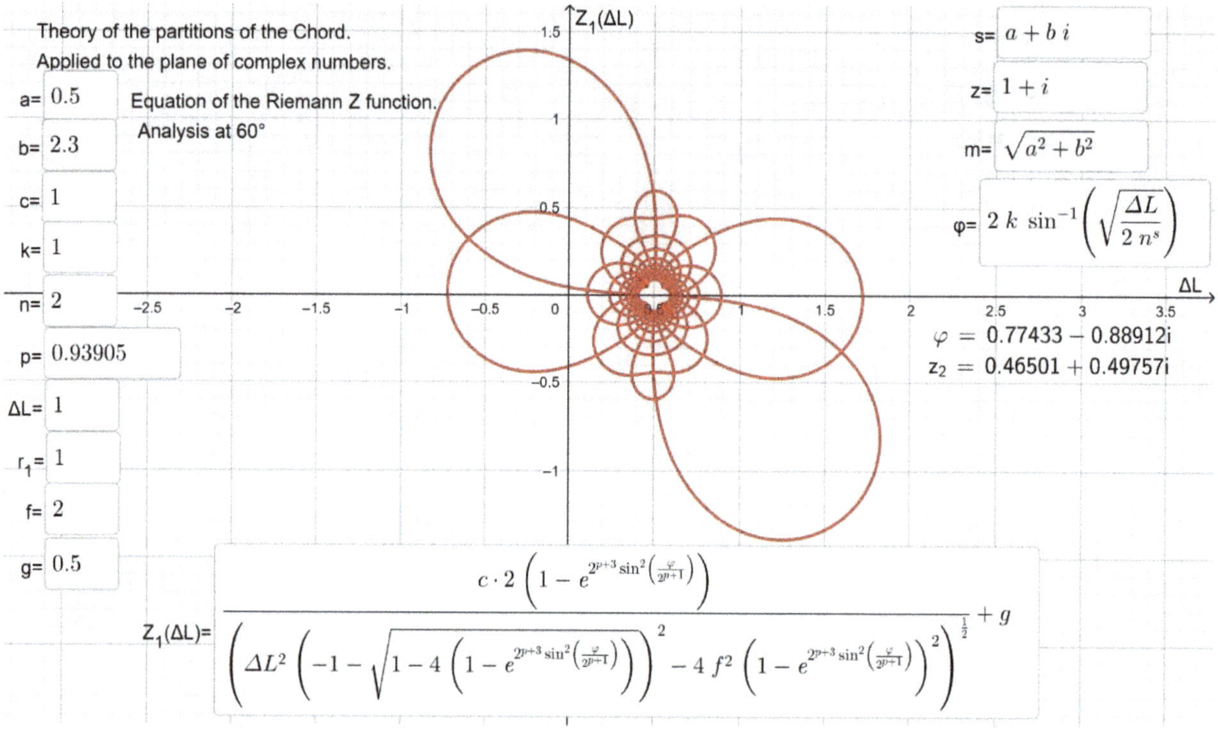

Figure 2.47: Riemann Z Function In The Plane Of Complex Numbers.

The following graphic is the application of the Trigonometric Partitions of the Chord to the Riemann Z Function, of the negative solution of the quadratic equation, with the equation of the spiral angle φ. In this analysis, the equation obtained for the Riemann Hypothesis at 60° is used. At the center of the Cartesian plane, with **b** and **f** varying in time.

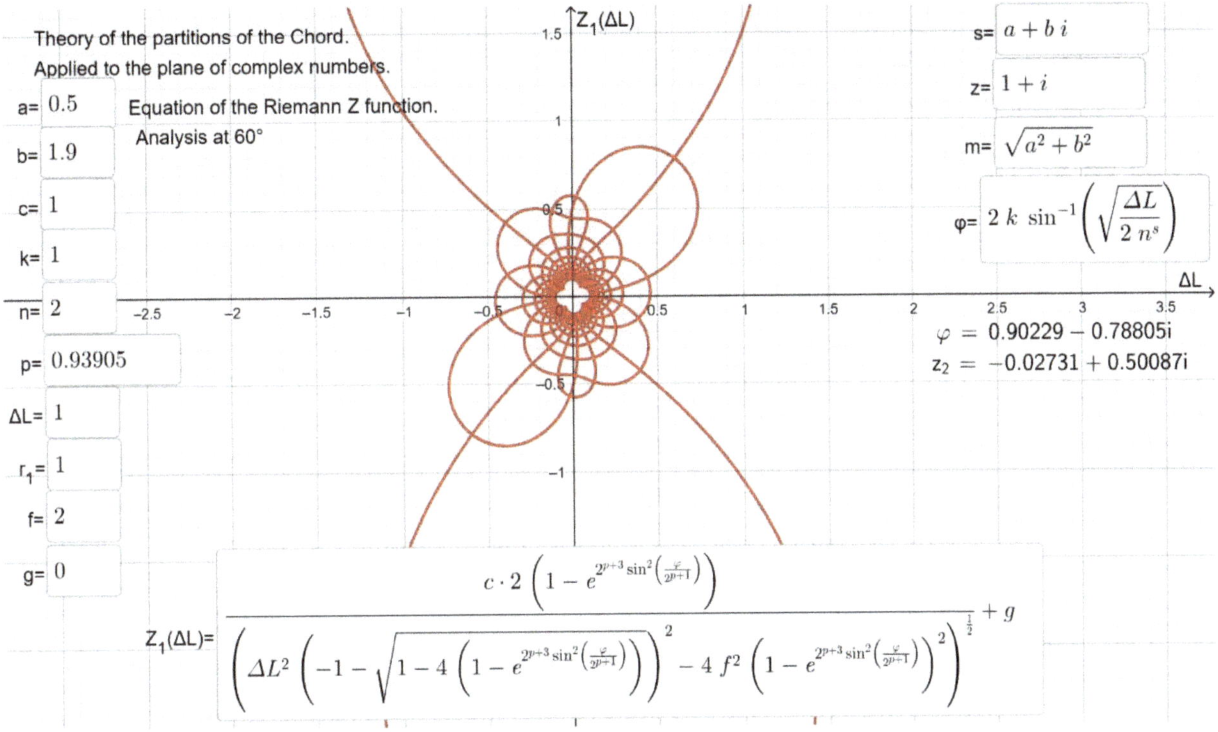

Figure 2.48: Riemann Z Function In The Plane Of Complex Numbers.

The following graphic is the application of the Trigonometric Partitions of the Chord to the Riemann Z Function, of the negative solution of the quadratic equation, with the equation of the spiral angle φ. In this analysis, the equation obtained for the Riemann Hypothesis at 60° is used. Moved on the **X** axis; with g=0.5 of the Cartesian plane, with **b** and **f** varying in time.

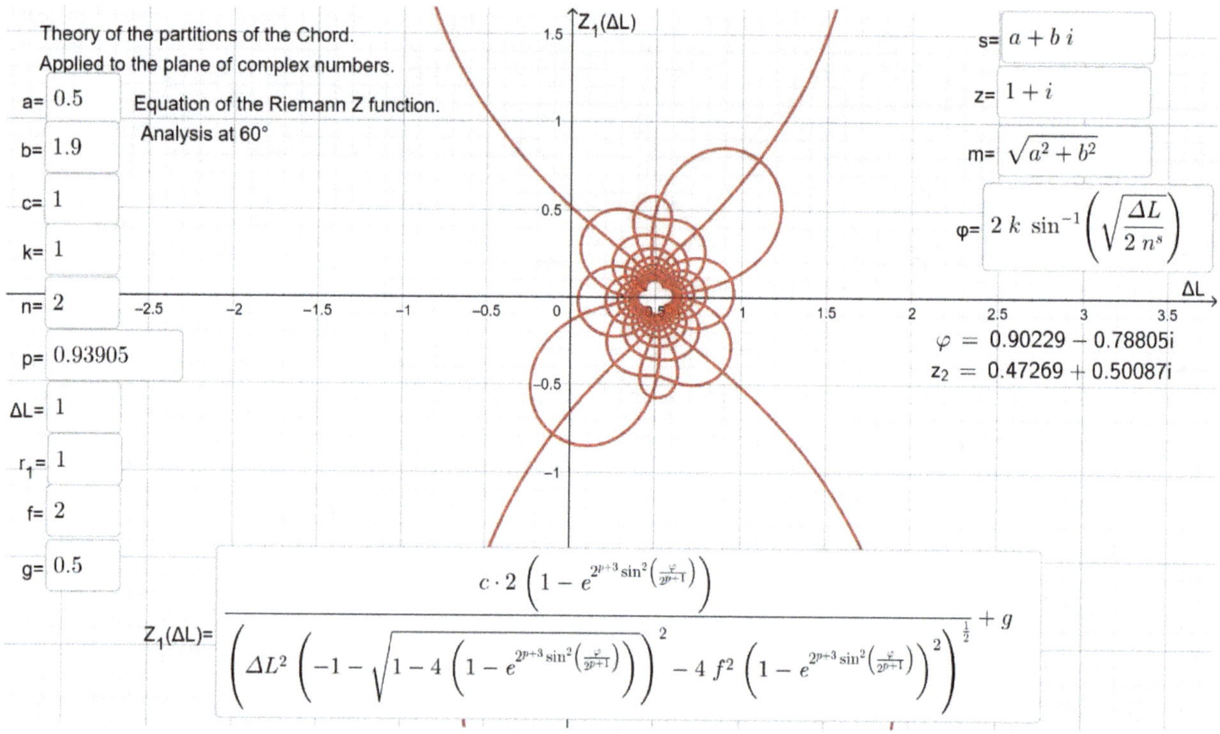

Figure 2.49: Riemann Z Function In The Plane Of Complex Numbers.

The following graphic is the application of the Trigonometric Partitions of the Chord to the Riemann Z Function, of the positive solution of the quadratic equation, with the equation of the spiral angle; $\varphi = 2 * k * \arcsin \sqrt{[\Delta L/(2 * n^s)]}$. In this analysis, the equation obtained for the Riemann Hypothesis at 60° is used. Located at the center of the Cartesian plane.

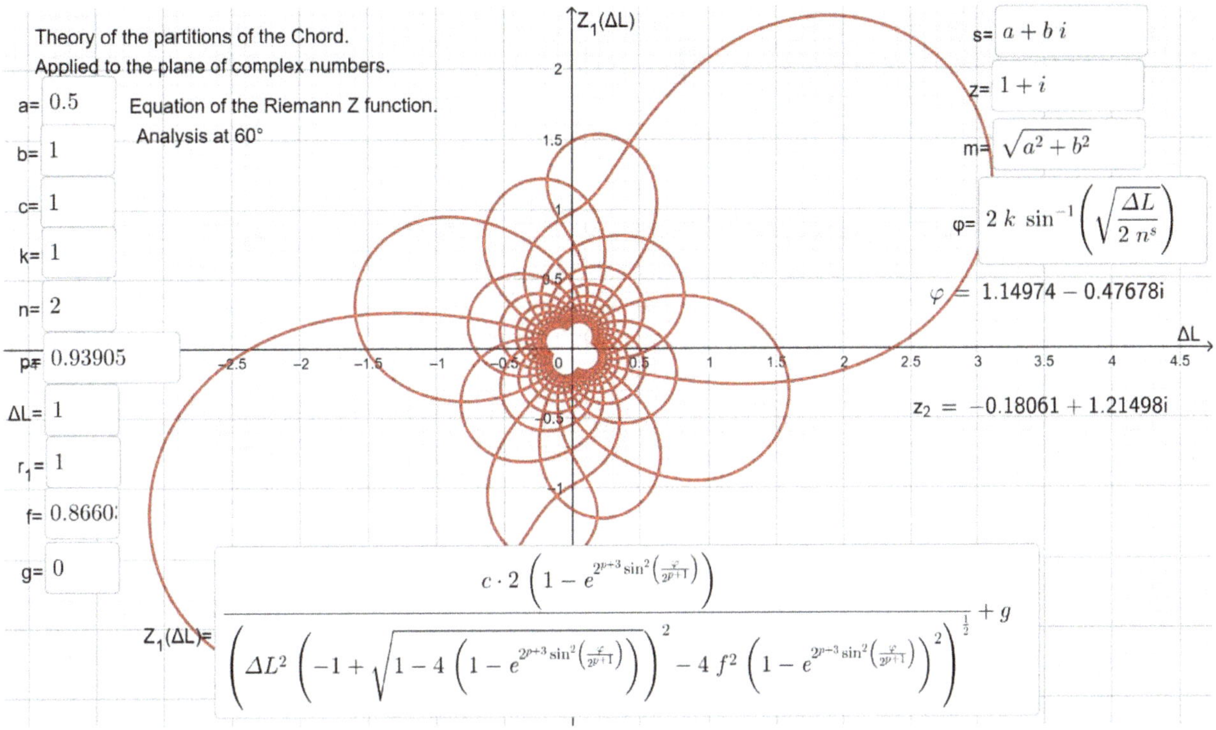

Figure 2.50: Riemann Z Function In The Plane Of Complex Numbers.

The following graphic is the application of the Trigonometric Partitions of the Chord to the Riemann Z Function, of the positive solution of the quadratic equation, with the equation of the spiral angle φ. In this analysis, the equation obtained for the Riemann Hypothesis at 60° is used. Moved on the X axis; with g=0.5. With these equations, the non-trivial zeros in the plane of complex numbers can be obtained. If s=a+bi, are constants, **c** which determines the size of the graph which is changing in time from minus infinity to plus infinity, we can determine the non-trivial zeros, **k** turns, **p** trigonometric partitions, ΔL the length module of the prime numbers, **n** is the subsequent real prime number; are not varying in time, the equations $+Z(\Delta L)$ and $-Z(\Delta L)$ generate the complex numbers Z_2,Z_3,Z_4 and Z_5 in GeoGebra where we will get all non-trivial zeros with the two equations. Likewise, you can place all the variables changing in time and you will get more non-trivial zeros.

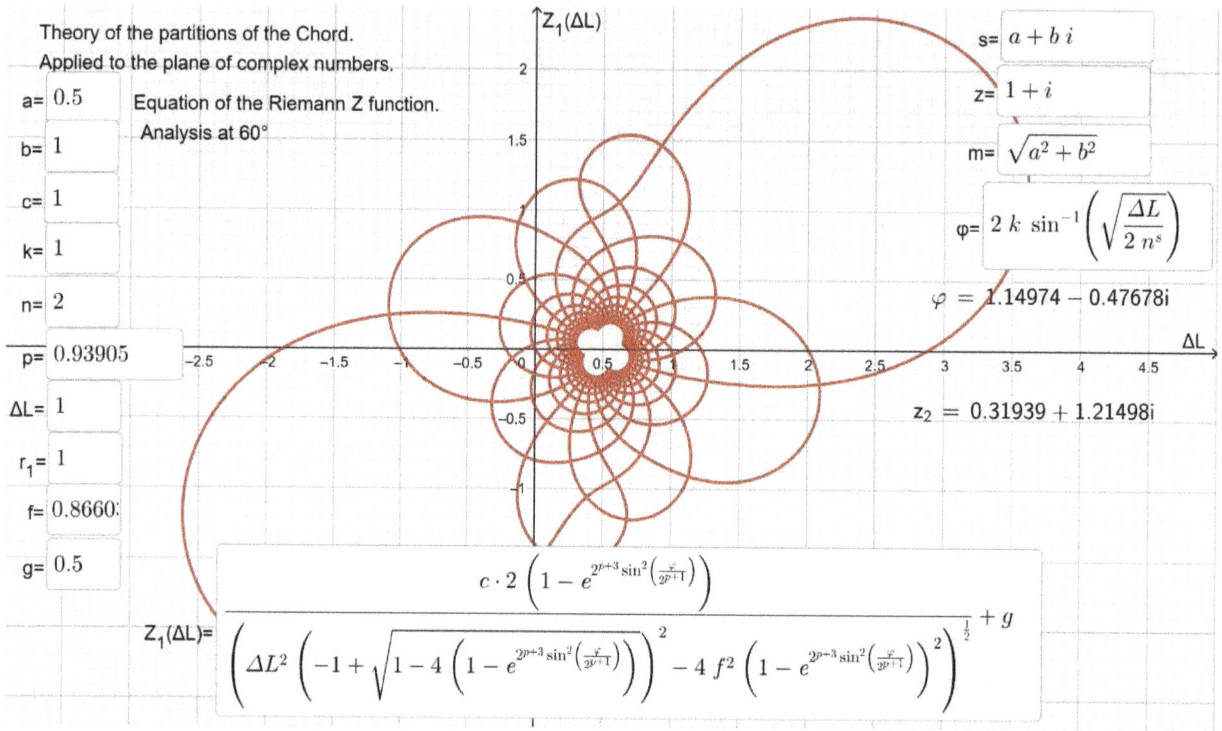

Figure 2.51: Riemann Z Function In The Plane Of Complex Numbers.

The following graphic is the application of the Trigonometric Partitions of the Chord to the Riemann Z Function, of the positive solution of the quadratic equation, with the equation of the spiral angle φ. In this analysis, the equation obtained for the Riemann Hypothesis at 60° is used. At the center of the Cartesian plane, with **b** and **f** varying in time.

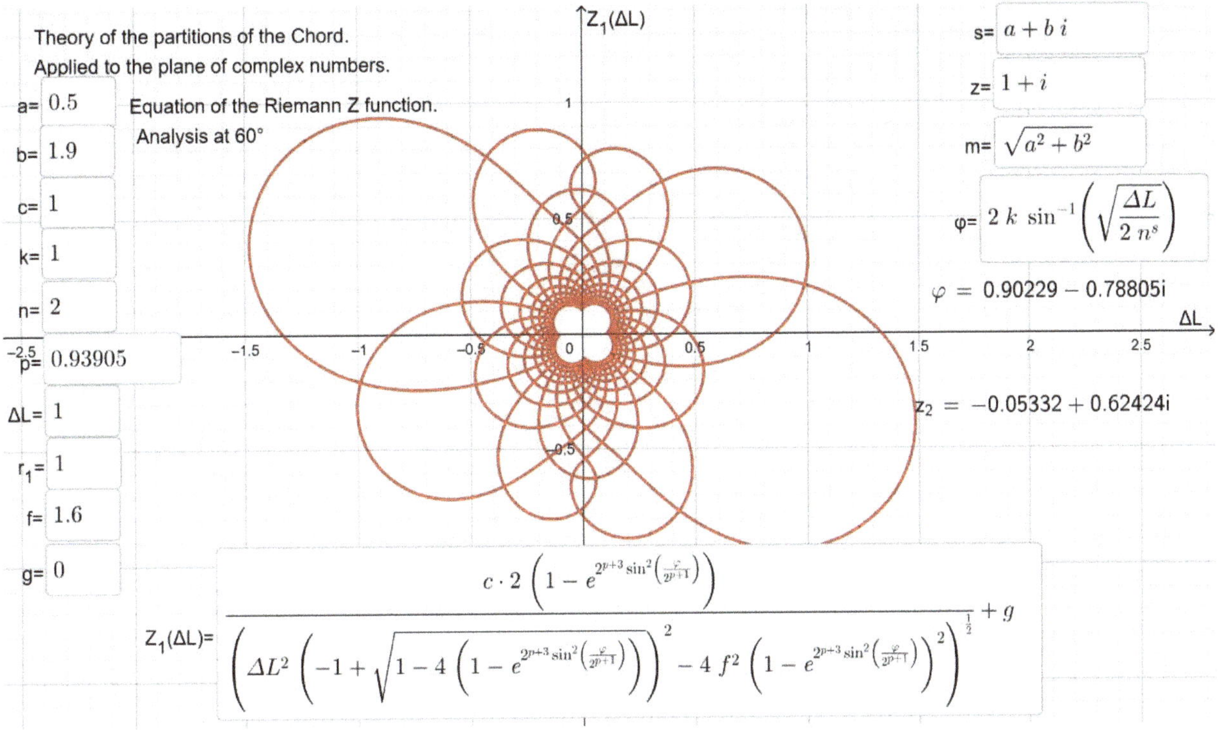

Figure 2.52: Riemann Z Function In The Plane Of Complex Numbers.

The following graphic is the application of the Trigonometric Partitions of the Chord to the Riemann Z Function, of the positive solution of the quadratic equation, with the equation of the spiral angle φ. In this analysis, the equation obtained for the Riemann Hypothesis at 60° is used. Moved on the **X** axis; with g=0.5 of the Cartesian plane, with **b** and **f** varying in time.

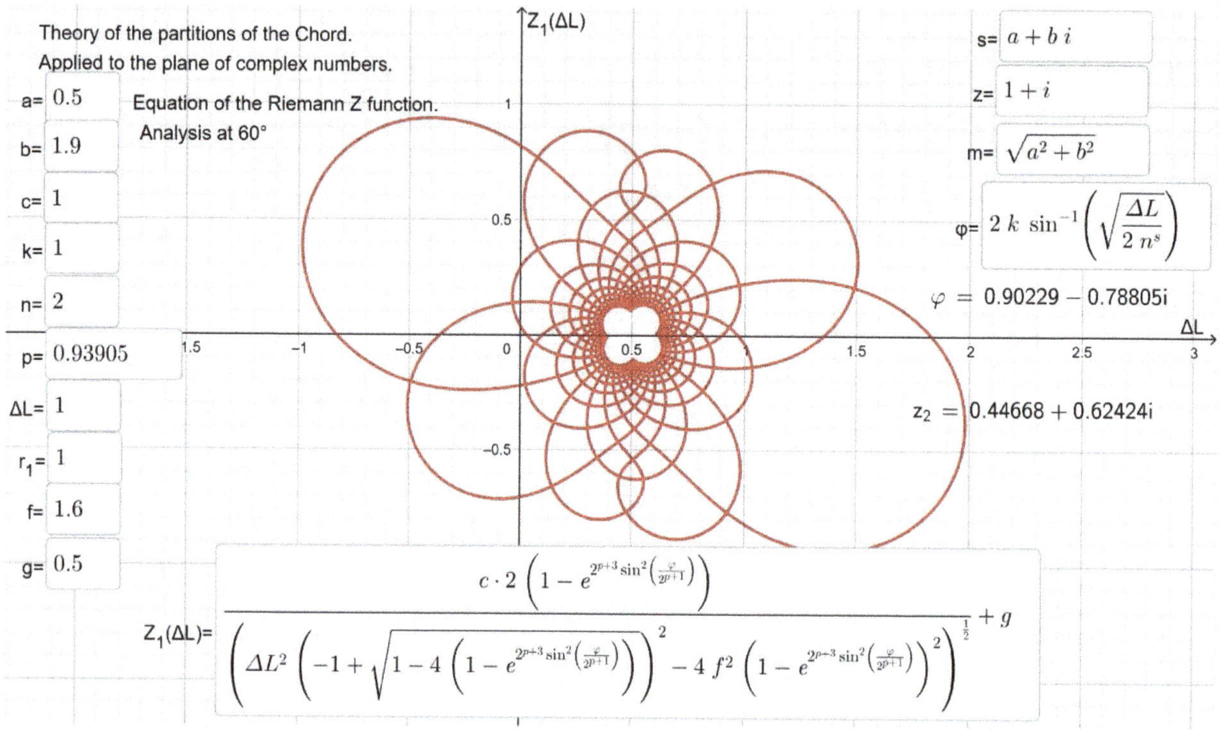

Figure 2.53: Riemann Z Function In The Plane Of Complex Numbers.

The following graphic is the application of the Trigonometric Partitions of the Chord to the Riemann Z Function, of the positive solution of the quadratic equation, with the equation of the spiral angle φ. In this analysis, the equation obtained for the Riemann Hypothesis at 60° is used. At the center of the Cartesian plane, with **b** and **f** varying in time.

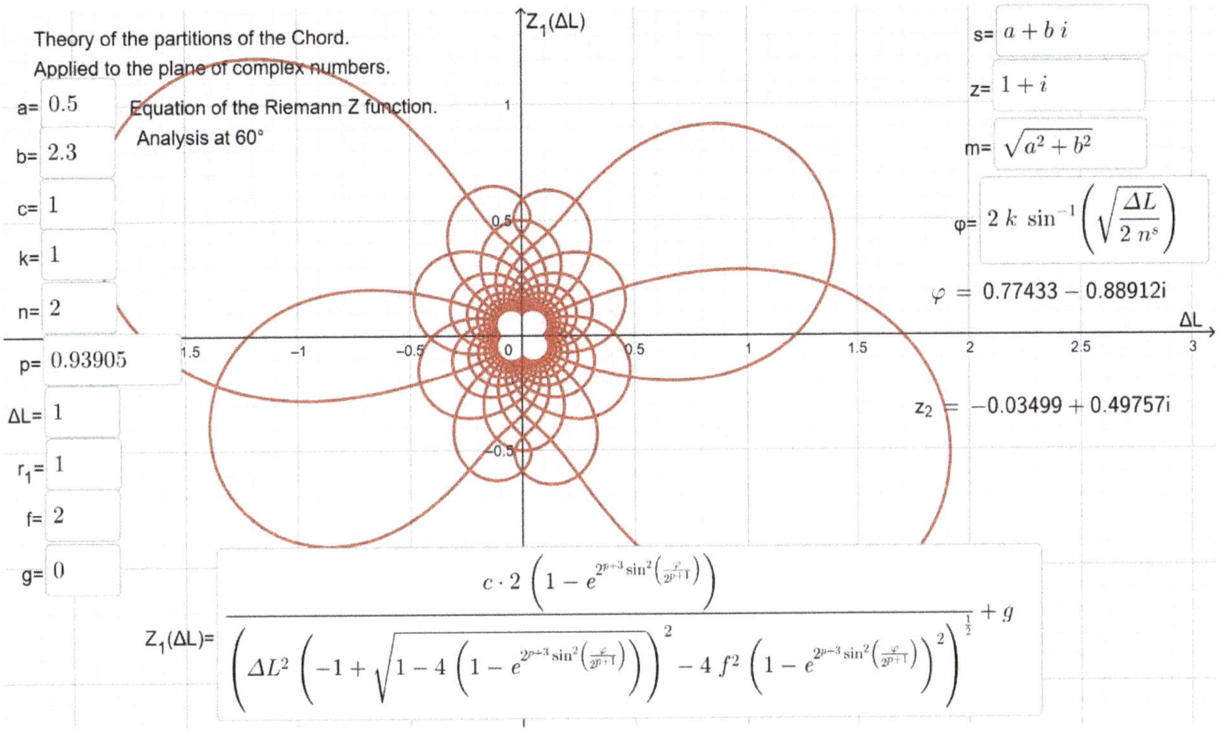

Figure 2.54: Riemann Z Function In The Plane Of Complex Numbers.

The following graphic is the application of the Trigonometric Partitions of the Chord to the Riemann Z Function, of the positive solution of the quadratic equation, with the equation of the spiral angle φ. In this analysis, the equation obtained for the Riemann Hypothesis at 60° is used. Moved on the **X** axis; with g=0.5 of the Cartesian plane, with **b** and **f** varying in time.

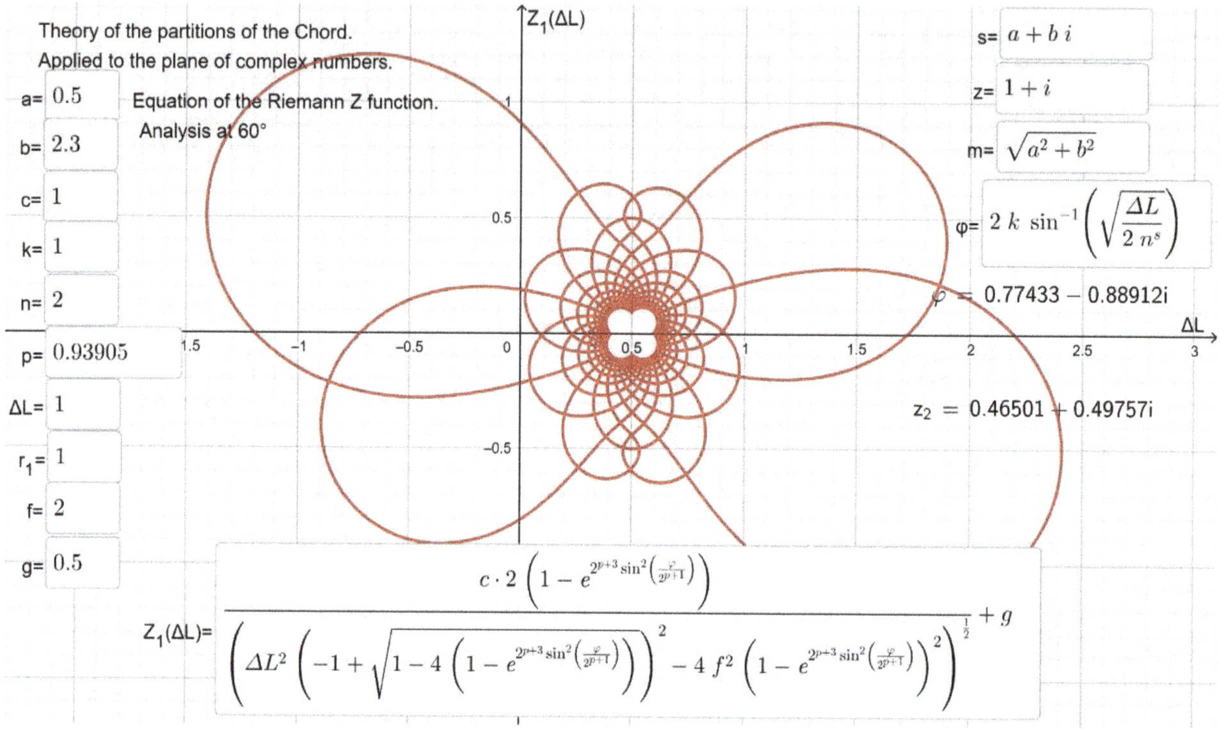

Figure 2.55: Riemann Z Function In The Plane Of Complex Numbers.

The following graphic is the application of the Trigonometric Partitions of the Chord to the Riemann Z Function, of the solution of the quadratic equation, with the equation of the spiral angle φ. In this analysis, the equation obtained for the Riemann Hypothesis at $90°$ is used. Moved on the X axis;

$Z(\Delta L) = \sum_{n=1}^{\infty}(1/n^s) = \sqrt{[e^{[2^{(p+3)}*\sin^2(\theta/2^{(p+1)})]} - 1]/[\Delta L^2 - f^2 * [e^{[2^{(p+3)}*\sin^2(\theta/2^{(p+1)})]} - 1]]}.$

with f=0,g=0.5

$Z(\Delta L) = \sum_{n=1}^{\infty}(1/n^s) = \sqrt{[e^{[2^{(p+3)}*\sin^2(\theta/2^{(p+1)})]} - 1]}/\Delta L.$

In General Form: $Z(\Delta L) = \sum_{n=1}^{\infty}(1/n^s) = c * \sqrt{[e^{[2^{(p+3)}*\sin^2(\theta/2^{(p+1)})]} - 1]}/\Delta L + g.$

This is one of the graphs that is equal to the Riemann Hypothesis, especially when analyzed on the unit circle, n=1.

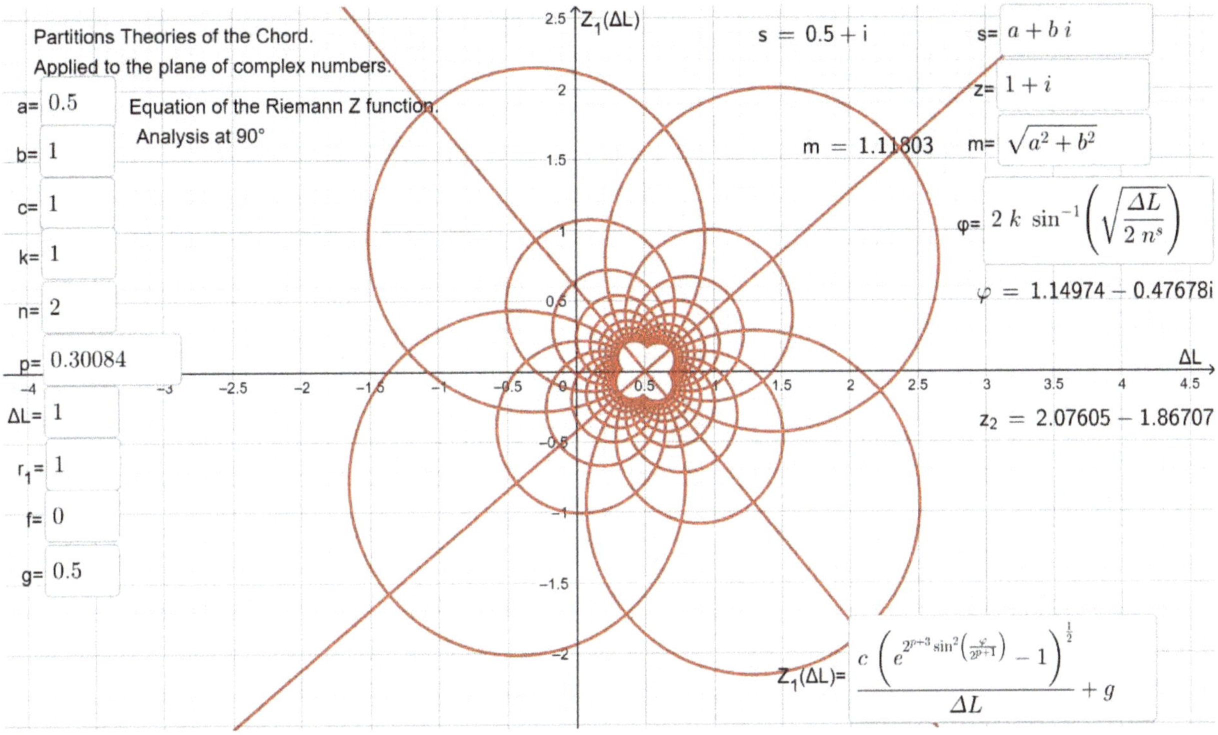

Figure 2.56: Riemann Z Function In The Plane Of Complex Numbers.

Each trivial and non-trivial zero can be reviewed and checked in the GeoGebra program where each formula was developed.

If we make the analogy of

$1/n^s = 1/P^s = e^{[-s*\ln(n)]} = e^{[-s*\ln(P)]}$; Where **n** is a real number and **P** is the Prime Number. Therefore, with n=P; $n^{(-\varphi*i)} = P^{(-\varphi*i)} = e^{(-\theta*i)} = \cos(\theta) - i*\sin(\theta)$.

Also, Moivre's theorem can be used to generalize the equation.

$n^{(-\varphi*x*i)} = P^{(-\varphi*x*i)} = e^{(-\theta*x*i)} = \cos(x*\theta) - i*\sin(x*\theta)$. Where x is a real or irrational number. With these equations, we can find that $\theta = \varphi*\ln(P)$. Therefore, $n^{(-\varphi*x*i)} = P^{(-\varphi*x*i)} = e^{(-\varphi*\ln(p)*x*i)} = \cos(x*\varphi*\ln(P)) - i*\sin(x*\varphi*\ln(P))$. With,$P = 2^L*\cot(\varphi)$; It is observed that Prime Numbers are numbers nested by levels that generate trigonometric spirals.

$\varphi = 2*k*\arcsin\sqrt{[\Delta L/(2*n^s)]}$ or

$\varphi = 2*k*\arcsin\sqrt{[\Delta r/(2*r^s)]}$ Finally, $\varphi = 2*k*\arcsin\sqrt{[(r_n - r_1)/(2*P^s)]}$.

$\varphi = 2*k*\arcsin\sqrt{[(\Delta P/2)*(1/P^s)]}$, $\varphi = 2*k*\arcsin\sqrt{[((P-P_0)/2)*(1/P^s)]}$.

The angle φ is what connects us with the twists and the prime numbers to the general equation for the Riemann Zeta function.

$\varphi = 2 * k * \arcsin\sqrt{[\Delta P/(2 * P^s)]} = \varphi = 2 * k * \arcsin\sqrt{[(P - P_0)/(2 * P^s)]}$.

If we solve for the variable P of the prime numbers; with $\Delta L = P - P_0$, and $P_0/P^s = 0$, Where P^s is so large compared to P_0, making this equation, as a function of **s** to the plane of the complex numbers, we will obtain a graph similar to that of the Riemann Z function. $P(s) = (0.5 * csc^2(\varphi/(2 * k))^{(s-1)^{-1}}$, . Likewise, P can be rewritten in terms of the angle $P(\varphi) = (0.5 * csc^2(\varphi/(2 * k))^{(s-1)^{-1}}$. Also, we can set $\Delta L = P^s - P_0^s$, we will obtain that P as a function, is equal to $P(s) = (sec(\varphi/(2 * k))^{-s} * P_0$, with $\varphi = 2^p * k * \arcsin\sqrt{[\Delta P/(2 * P^s)]}$ in terms of the trigonometric partitions and $P(\varphi) = (sec(\varphi/(2 * k))^{-s} * P_0$; this tells us the patterns that can generate prime numbers.

$Z(\Delta L) = \sum_{n=1}^{\infty}(1/n^s) =$
$\sqrt{[e^{[2^{(p+3)} * \sin^2(2*k*\arcsin\sqrt{[(\Delta L/2)*(1/n^s)]}/2^{(p+1)})]} - 1]/\Delta L}$.Either

$Z(\Delta r) = \sum_{n=1}^{\infty}(1/n^s) =$
$\sqrt{[e^{[2^{(p+3)} * \sin^2(2*k*\arcsin\sqrt{[(\Delta r/2)*(1/r^s)]}/2^{(p+1)})]} - 1]/\Delta r}$.

$Z(\Delta P) = \sum_{n=1}^{\infty}(1/n^s) =$
$\sqrt{[e^{[2^{(p+3)} * \sin^2(2*k*\arcsin\sqrt{[(\Delta P/2)*(1/P^s)]}/2^{(p+1)})]} - 1]/\Delta P}$.

If we consider some values of non-trivial zeros, already found by great mathematicians such as Friedrich Bernhard Riemann and Carl Ludwig Siegel: And we introduce them into the equation:

$Z(\Delta L) = \sum_{n=1}^{\infty}(1/n^s) = \sqrt{[e^{[2^{(p+3)} * \sin^2(\varphi/2^{(p+1)})]} - 1]/\Delta L}$.

Where n=1, with $\Delta L = \Delta r$, n different from zero, and $\varphi = 2 * k * \arcsin\sqrt{[\Delta L/(2 * n^s)]}$

For example: $Z(\Delta L) = \sum_{n=1}^{\infty}(1/n^s) =$
$\sqrt{[e^{[2^{(p+3)} * \sin^2(2*k*\arcsin\sqrt{[\Delta L/(2*n^s)]}/2^{(p+1)})]} - 1]/\Delta L}$.

Entering the values from this table of Friedrich Bernhard Riemann and Carl Ludwig Siegel into $Z(\Delta L)$ equation we will obtain the following graph that remains the same without changes.

Table 2.1: Values of the non-trivial zeros of the critical band

Values of the non-trivial zeros of the critical band. S=a+bi

Quantity	a	b	Complexes(s)	Radii Growths Δr	Modulus(r)	Arguments(φ)	Arguments($\varphi°$)
1	0.5	14.13472531	0.5+14.13472531i	13.64356601	14.1436	1.5354372	87.974071
2	0.5	21.02203964	0.5+21.0220396387715i	20.52798494	21.028	1.5470162	88.637502
3	0.5	25.01085758	0.5+25.0108575801456i	24.51585491	25.0159	1.5508077	88.854734
4	0.5	30.42487613	0.5+30.4248761258595i	29.92898433	30.429	1.5543639	89.05849
5	0.5	32.93506159	0.5+32.9350615877391i	32.43885672	32.9389	1.5556161	89.130237
6	0.5	37.58617816	0.5+37.5861781588256i	37.0895037	37.5895	1.5574943	89.237853
7	0.5	40.91871901	0.5+40.9187190121474i	40.42177373	40.9218	1.5585776	89.299918
8	0.5	43.32707328	0.5+43.3270732809149i	42.82995822	43.33	1.5592567	89.338829
9	0.5	48.00515088	0.5+48.0051508811671i	47.5077547	48.0078	1.5603812	89.403255
10	0.5	49.77383248	0.5+49.7738324776723i	49.27634377	49.7763	1.5607512	89.424458
11	0.5	56.44624770	0.5+56.4462476970633i	55.94846215	56.4485	1.5619386	89.492488
12	0.5	59.34704400	0.5+59.3470440026023i	58.84915022	59.3492	1.5623715	89.517293

If we vary the rest of the variables in this equation, the graph that is generated is different in size and position. In the same way, with the variation of the variables, its shape is maintained. The following graph is the application of the Trigonometric Partitions of the Chord to the Riemann Z Function, of the solution of the quadratic equation, with the equation of the spiral angle φ for circular waves.In this analysis the equation that was obtained for the Riemann Hypothesis at 90° is used. Moved in the X axis; with a=0.5,

b= 14.13472531, g=0.5, n=1.

Remembering that $\Delta L = \Delta r = \Delta P = P - P_0$; Posterior Prime Number minus Previous Prime Number.

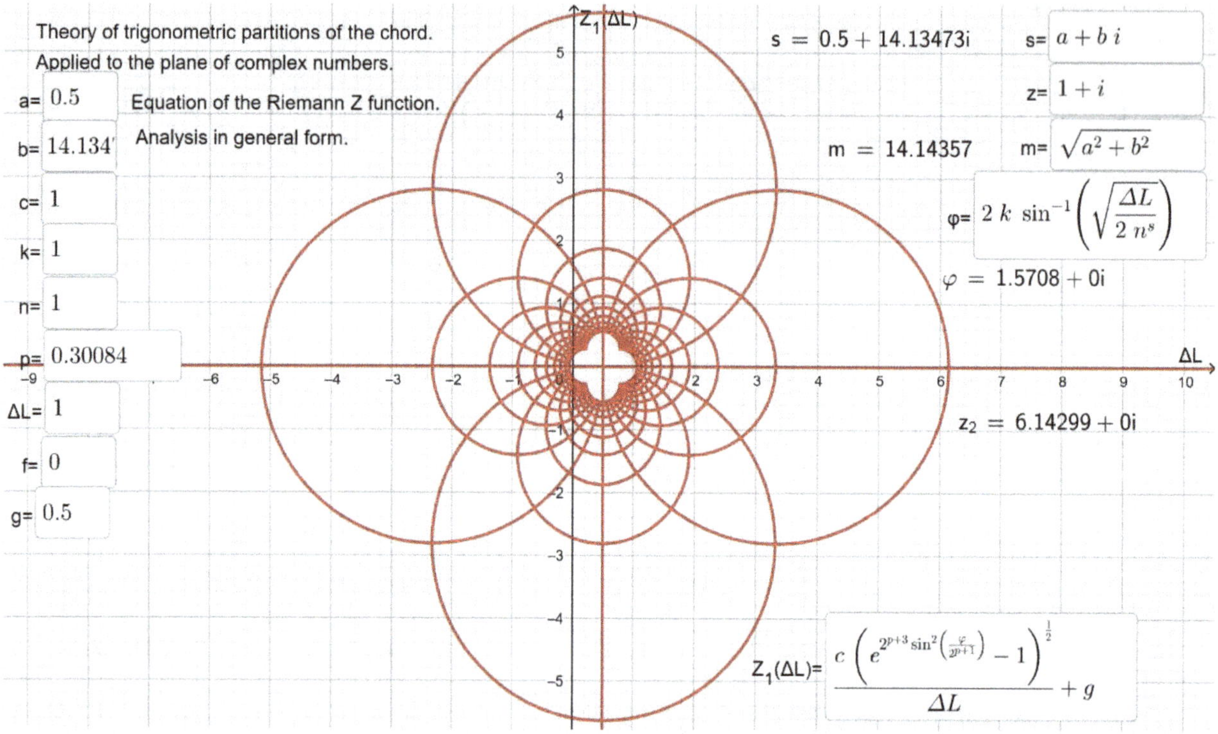

Figure 2.57: Riemann Z Function In The Plane Of Complex Numbers.

In the final section: Analysis of the Riemann Zeta Function in Relation to Prime Numbers, I demonstrate that all the equations of trigonometric partitions can be used to generate the graph of the Riemann Zeta function in its general form. Likewise, these same formulas can be written in a simplified form with $f = 0$ and $\varphi_0 = \pi/2$, for example: Trigonometric Partition equation for the growth of the radius:

$Z(\Delta L) = \sum_{n=1}^{\infty}(1/n^s) = c * \sqrt{[e^{[2*\varphi-2^p*\sin(\varphi/2^{(p-1)})]}-1]/\Delta L}.$

Trigonometric Partition equation for the arrow:

$Z(\Delta L) = \sum_{n=1}^{\infty}(1/n^s) = c * \sqrt{[e^{[2*\varphi-2^{(p+1)}*\sin(\varphi/2^p)]}-1]/\Delta L}.$

Trigonometric Partition equation for the apothem:

$Z(\Delta L) = \sum_{n=1}^{\infty}(1/n^s) = c * \sqrt{[e^{[2^{(p+1)}*\sin(\varphi/2^p)]}-1]/\Delta L}.$

The following graph is the application of the Trigonometric Partitions of the Chord to the Riemann Z Function, of the solution of the quadratic equation, with the equation of the spiral angle φ. In this analysis the equation that was obtained for the Riemann Hypothesis is at 90° is used. Moved in the X axis; with a=0.5, b= 21.0220396387715, f=0. g=0.5, n=2.

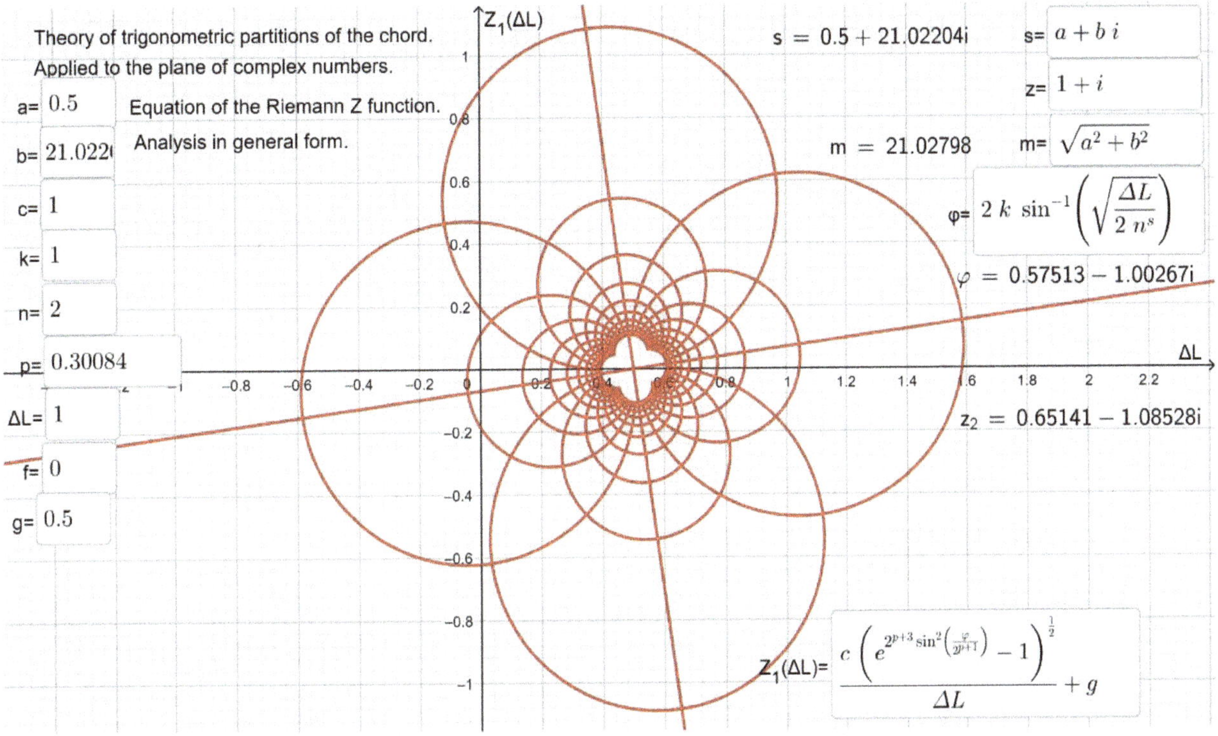

Figure 2.58: Riemann Z Function In The Plane Of Complex Numbers.

We can use this same table of complex numbers in the equation and introduce them into the equation: $Z(\Delta L) = \sum_{n=1}^{\infty}(1/n^s) = \sqrt{[e^{[2^{(p+3)}*\sin^2(\varphi/2^{(p+1)})]} - 1]/\Delta L}$.

Where n=1, n different from zero, with $\Delta L = \Delta r = r_n - r_1$, and $\varphi = 2*k*\arcsin\sqrt{[\Delta L/(2*n^s)]}$.

For example: $Z(\Delta L) = \sum_{n=1}^{\infty}(1/n^s) =$

$\sqrt{[e^{[2^{(p+3)}*\sin^2(2*k*\arcsin\sqrt{[\Delta L/(2*n^s)]}/2^{(p+1)})]} - 1]/\Delta L}$.

General Form:

$Z(\Delta L) = \sum_{n=1}^{\infty}(1/n^s) =$

$c * \sqrt{[e^{[2^{(p+3)}*\sin^2(2*k*\arcsin\sqrt{[\Delta L/(2*n^s)]}/2^{(p+1)})]} - 1]/\Delta L} + g$.

Entering the values from this table into $Z(\Delta L)$ equation we will obtain the following graph that remains the same without changes. In the book of Spiral Angles, Spirals and Trigonometric Partitions there is the following table that was obtained with a basic equation of Prime Numbers where **P**: Is the prime number and **L** is the level where the prime number is located. $P = 2^L * \cot(\varphi)$ where **L** is the level, it is greater than and equal to zero. With this equation the phi angle φ for circular waves was obtained.

The following graph is the application of the Trigonometric Partitions of the Chord to the

Table 2.2: Values of the non-trivial zeros of the critical band

L	#	$\Delta r = \Delta P$	P	2^L	φ°	$P \cdot \cos(\varphi)$	$P \cdot \sin(\varphi)$	Complexes(s=a+bi)
0	1	1	1	1	45.00000000000	0.70710678119	0.70710678119	0.707106781186548+0.707106781186547i
0	2	1	2	1	26.56505117708	1.78885438200	0.89442719100	1.78885438199983+0.894427190999916i
0	3	1	3	1	18.43494882292	2.84604989415	0.94868329805	2.84604989415154+0.948683298050514i
1	**4**	**2**	**5**	**2**	**21.80140948635**	**4.64238345443**	**1.85695338177**	**4.6423834544263+1.85695338177052i**
2	5	2	7	4	29.74488129694	6.07770199487	3.47297256850	6.07770199487121+3.47297256849784i
3	6	4	11	8	36.02737338510	8.89609692734	6.46988867443	8.89609692733507+6.46988867442551i
4	7	**2**	13	16	50.90614111377	8.19770412623	10.08948200151	8.19770412622805+10.0894820015115i
5	8	4	17	32	62.02052561152	7.97563884135	15.01296723078	7.97563884134974+15.012967230776i
6	9	2	19	64	73.46516214265	5.40736748010	18.21429045928	5.40736748009989+18.2142904592838i
7	10	6	23	128	79.81337024006	4.06766658283	22.63744880880	4.06766658283078+22.6374488087974i
8	11	6	29	256	83.53701422968	3.26427835898	28.81569861716	3.26427835897542+28.8156986171622i
9	12	2	31	512	86.53514942840	1.87352217151	30.94333392950	1.873522171513+30.9433339295049i
10	13	6	37	1024	87.93064259913	1.33604219191	36.97587039221	1.336042191906+36.9758703922091i
11	14	4	41	2048	88.85311855151	0.82063634974	40.99178644535	0.820636349736082+40.9917864453538i
12	**15**	**2**	**43**	**4096**	**89.39852831614**	**0.45139114263**	**42.99763070259**	**0.451391142629728+42.9976307025905i**

Riemann Z Function, of the solution of the quadratic equation, with the equation of the spiral angle φ for circular waves. In this analysis the equation that was obtained for the Riemann Hypothesis at 90° is used. Moved in the X axis; with a=0.451391142629728, b= 42.9976307025905, m=43=P, f=0, g=0.5, n=1.

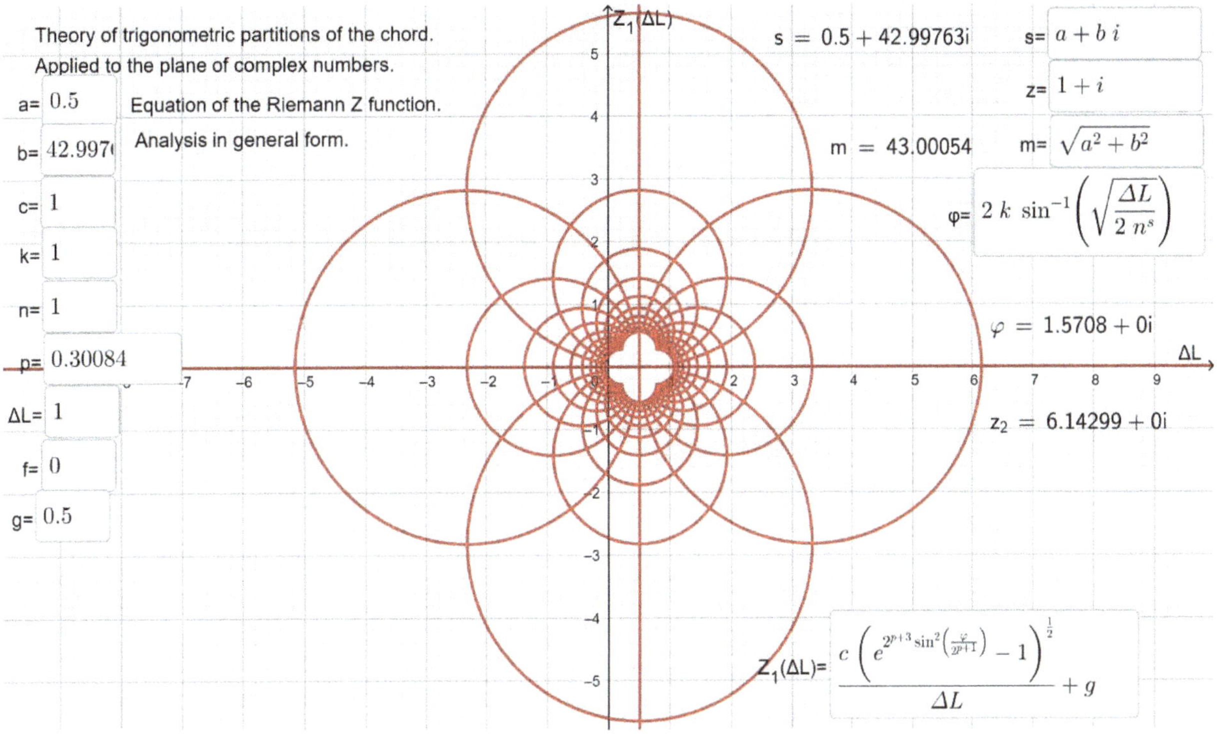

Figure 2.59: Riemann Z Function In The Plane Of Complex Numbers.

If we are in the complex number plane and substitute these mathematically equivalent angle equations, it leads us to the Cartesian plane of real numbers. With this equation, we can generate graphs of hyperbolic functions. If we use these equivalent angle equations in the equation of Figure 2.59, a graph of the hyperbolic function can be generated, with a horizontal asymptote equal to 1/2, and a vertical asymptote equal to zero.

With $\varphi = 2\pi n = \omega t = 2\pi f t = kx - \omega t$, where $k = \frac{2\pi}{\lambda}$, we can perform an analysis in the Cartesian plane of real numbers as a function of angular velocity and time, frequency and time, position and time, the wave number k, and λ, its wavelength.That is the reason why, in some formulas, the graphs are generated in two halves due to their two asymptotes, as seen in this graph.

Each trivial and non-trivial zero can be obtained, reviewed and checked in the GeoGebra program where each formula was developed and that the Riemann hypothesis is true.

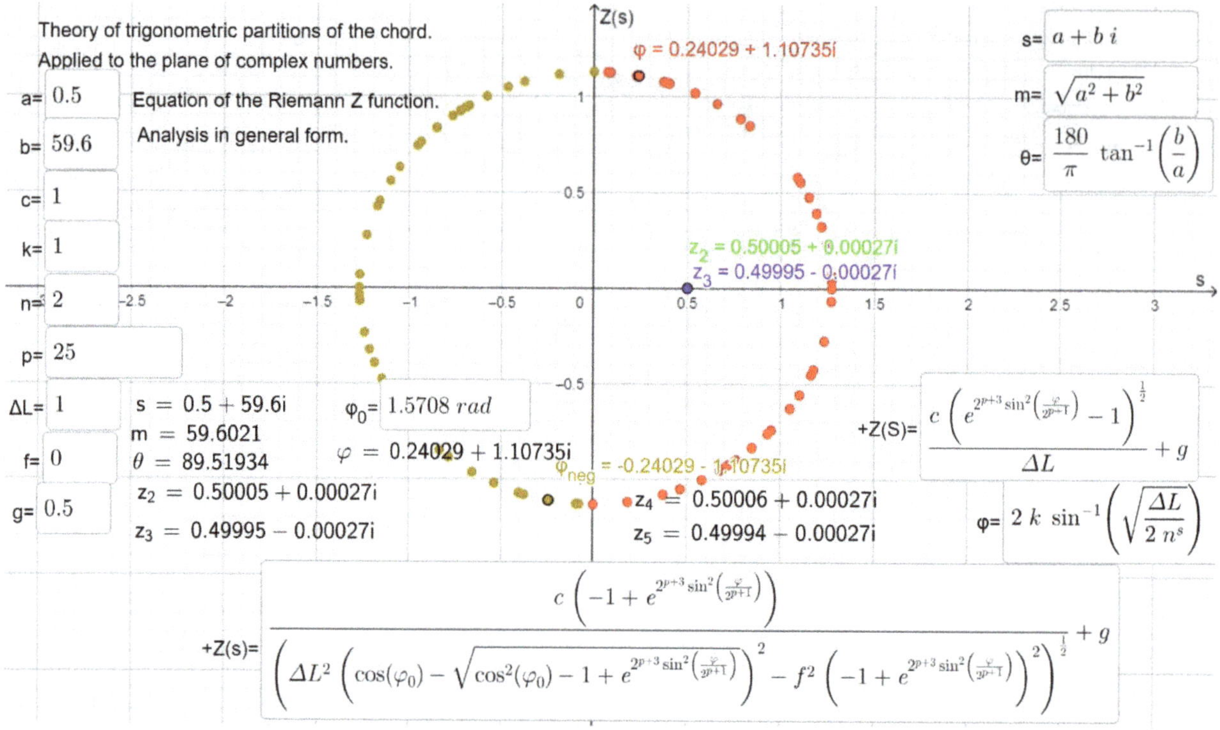

Figure 2.60: Dots Plot of the Riemann Z(s) Function.

Finally, the equation of the Riemann Z function can be written in its general form: Positive quadratic solution: The graph in the complex plane will be composed of two halves, one positive $+Z(\Delta L)_1$ and the other negative $-Z(\Delta L)_2$. Where φ_0 is the initial angle of the analysis,The angle phi sub zero is for rotating the second axis parallel to the X-axis. That's why when the graph of the Riemann Zeta function is formed, additional axes appear in the figure.**C** is a constant to make the graph bigger, **g** is a constant to move the graph along the x-axis.φ is the angle to which the trigonometric partitions are applied.And **p** are the trigonometric partitions of the angle.

It is a transformation of the Riemann Zeta Function;because we are changing from the plane of real numbers to the plane of complex numbers with the same equations.

Z(ΔL) In the plane is the graph of circles in the complex number plane and reveals the graph of the Riemann Zeta Function Z(ΔL)=$\sum_{n=1}^{\infty}(1/n^s)$= ΔL=Δr=$r_n - r_1$.
$Z(s)$ is the graph of dots in the polar Cartesian plane and complex numbers depending on which equation is being used; for example, when the Riemann equation for 90° degrees is used. In some cases, this equation does not generate graphs, it is a dot or invisible; for example, when the general formula of the Riemann Zeta function is used. Remembering that; $Z(\Delta L) \neq$Z(s), $\Delta L = \Delta r = r_n - r_1 = P_n - P_0$.
The Riemann Zeta function of the summation of the inverse of real numbers raised to the power of S, with n equal to one up to infinity; which can be for real or imaginary numbers; is the equation that generates dots in the complex number plane. The equation of the Euler prime product in terms of prime numbers with their power s; is the equation that generates flat circular waves in the complex number plane, it is the one that reveals the graph of the Riemann Hypothesis. Meanwhile, the Riemann Zeta s function is the one that generates dots as a graph, and the one that helps us calculate the trivial zeros in the real plane and the non-trivial zeros in the imaginary plane. With these equations of trigonometric partitions, trial, and error can be used to calculate the values of the tables of trivial and non-trivial zeros discovered by Riemann and Siegel. Or if we use these values from those tables of Riemann and Siegel in the equations of trigonometric partitions, we will observe that the graph passes through those dots. Similarly, the tables with the trigonometric partitions developed in this text can be used to calculate all the trivial and non-trivial zeros, or a combination of the Riemann-Siegel tables and the Trigonometric Partitions with which the non-trivial zeros can be calculated in the complex number plane and the trivial zeros in the real plane.In these equations involving Z(s) and $Z(\Delta L)$ with variables b and n, we can use prime numbers to perform their calculations, and we will observe that their values are equal or oscillate around the real number 1/2.(Edwards, 2001)

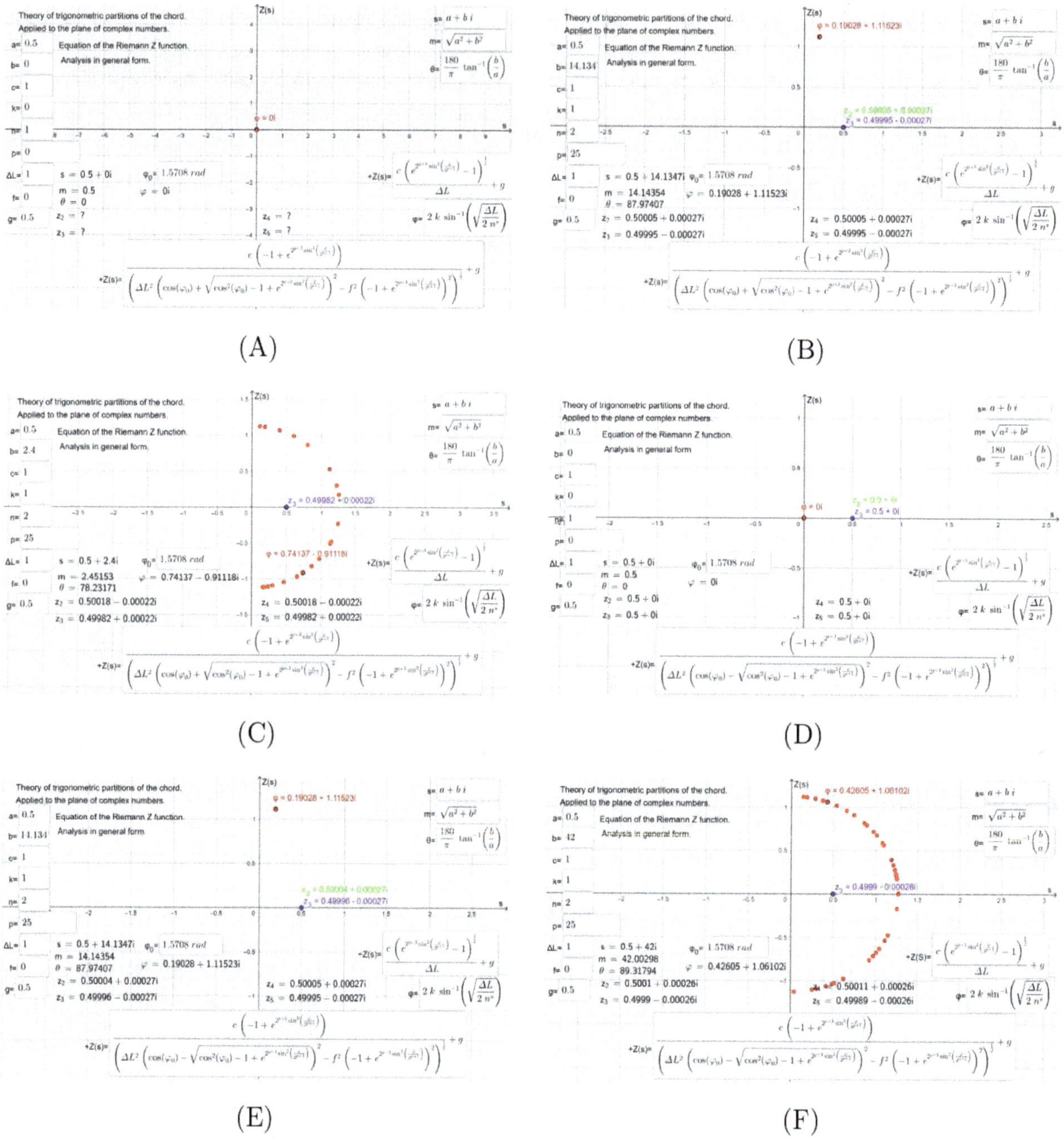

Figure 2.61: Dots Plots of the Riemann Z(s) Function.

Positive quadratic solution: The graph in the complex plane will be composed of two halves, one positive $+Z(\Delta L)_2$ and the other negative $-Z(\Delta L)_3$

$$Z(\Delta L) = \sum_{n=1}^{\infty}(1/n^s) = c * [e^{2^{(p+3)} * \sin^2(\varphi/2^{p+1})} - 1] \Big/ \sqrt{\Delta L^2 * [\cos(\varphi_0) + \sqrt{[\cos^2(\varphi_0) + e^{2^{(p+3)} * \sin^2(\varphi/2^{p+1})} - 1]}]^2 - f^2 * [e^{2^{(p+3)} * \sin^2(\varphi/2^{p+1})} - 1]^2 + g}.$$

Square root solution, positive:

$$Z(\Delta L) = \sum_{n=1}^{\infty}(1/n^s) = -c * [e^{2^{(p+3)} * \sin^2(\varphi/2^{p+1})} - 1] \Big/ \sqrt{\Delta L^2 * [\cos(\varphi_0) + \sqrt{[\sin^2(\varphi_0) + e^{2^{(p+3)} * \sin^2(\varphi/2^{p+1})}]}]^2 - f^2 * [e^{2^{(p+3)} * \sin^2(\varphi/2^{p+1})} - 1]^2 + g}.$$

Negative quadratic solution: The graph in the complex plane will be composed of two halves, one positive $+Z(\Delta L)_2$ and the other negative $-Z(\Delta L)_3$

$$Z(\Delta L) = \sum_{n=1}^{\infty}(1/n^s) = c * [e^{2^{(p+3)} * \sin^2(\varphi/2^{p+1})} - 1] \Big/ \sqrt{\Delta L^2 * [\cos(\varphi_0) - \sqrt{[\cos^2(\varphi_0) + e^{2^{(p+3)} * \sin^2(\varphi/2^{p+1})} - 1]}]^2 - f^2 * [e^{2^{(p+3)} * \sin^2(\varphi/2^{p+1})} - 1]^2 + g}.$$

Square root solution, negative:

$$Z(\Delta L) = \sum_{n=1}^{\infty}(1/n^s) = -c * [e^{2^{(p+3)} * \sin^2(\varphi/2^{p+1})} - 1] \Big/ \sqrt{\Delta L^2 * [\cos(\varphi_0) - \sqrt{[\sin^2(\varphi_0) + e^{2^{(p+3)} * \sin^2(\varphi/2^{p+1})}]}]^2 - f^2 * [e^{2^{(p+3)} * \sin^2(\varphi/2^{p+1})} - 1]^2 + g}.$$

These equations and graphs are super symmetric.At 90° degrees both equations give the same results. And also at 90° degrees both graphs are the same.In the positive solution of the equation; when analyzed with the angle phi sub zero φ_0 initials from 0° to 180° degrees the graph increases and from 180° to 360° degrees the graph decreases.The angle phi sub zero is for rotating the second axis parallel to the X-axis. That's why when the graph of the Riemann Zeta function is formed, additional axes appear in the figure. But it is the opposite of the negative solution of the equation.In these same equations we can use real or complex numbers to make the similar or same graphs.

Analyzing this equation on the unit circle, with a=0, b=0, s=0+0i, p=0, f=0, g=0.5, and the rest of the variables equal to one. With φ_0 in the range from 0 to π and from π to $2*\pi$; rotations and inversions of secondary axes parallel to x' and y' occur. For example, in this equation in the range from 0 to π, the graph of the Riemann Z function increases from its minimum size to maximum. From π to $2*\pi$, the graph decreases from its maximum size to its minimum.If the angle φ or φ_0 is positive or negative in its equation, the graph remains the same.

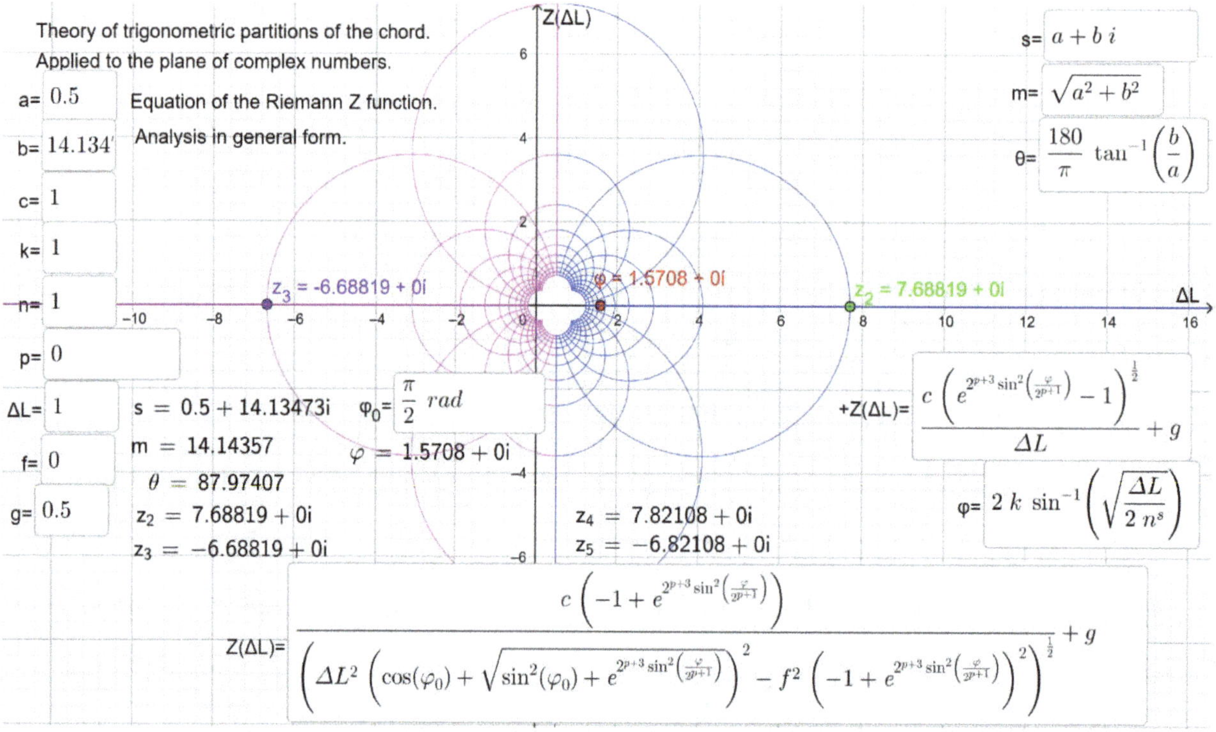

Figure 2.62: Riemann Z Function In The Plane Of Complex Numbers.

"In the general solution of the Riemann Z function; positive and negative with φ_0 equal to $\pi/2$ and $3*\pi/2$, for both positive and negative equations, their graphs of the Riemann Z function are identical."

With these equations from this section of the general solution of the Riemann Z function, we can analyze the concentric circles from the center, starting with an initial radius of zero, up to a number **n**, which can be any number or only prime numbers.

Where, $r_1 = P_0 = 0$, and $r_n = n$, $\Delta L = \Delta r = r_n - r_1 = \Delta P = P_n - P_0$.

Analyzing this equation on the unit circle, with a=0, b=0, s=0+0i, p=0, f=0, g=0.5, and the rest of the variables equal to one. With φ_0 in the range from 0 to π and from π to $2*\pi$; rotations and inversions of secondary axes parallel to x' and y' occur. For example, in this equation in the range from 0 to π, the graph of the Riemann Z function decreases from its maximum size to minimum. From π to $2*\pi$, the graph increases from its minimum size to its maximum.If the angle φ or φ_0 is positive or negative in its equation, the graph remains the same.

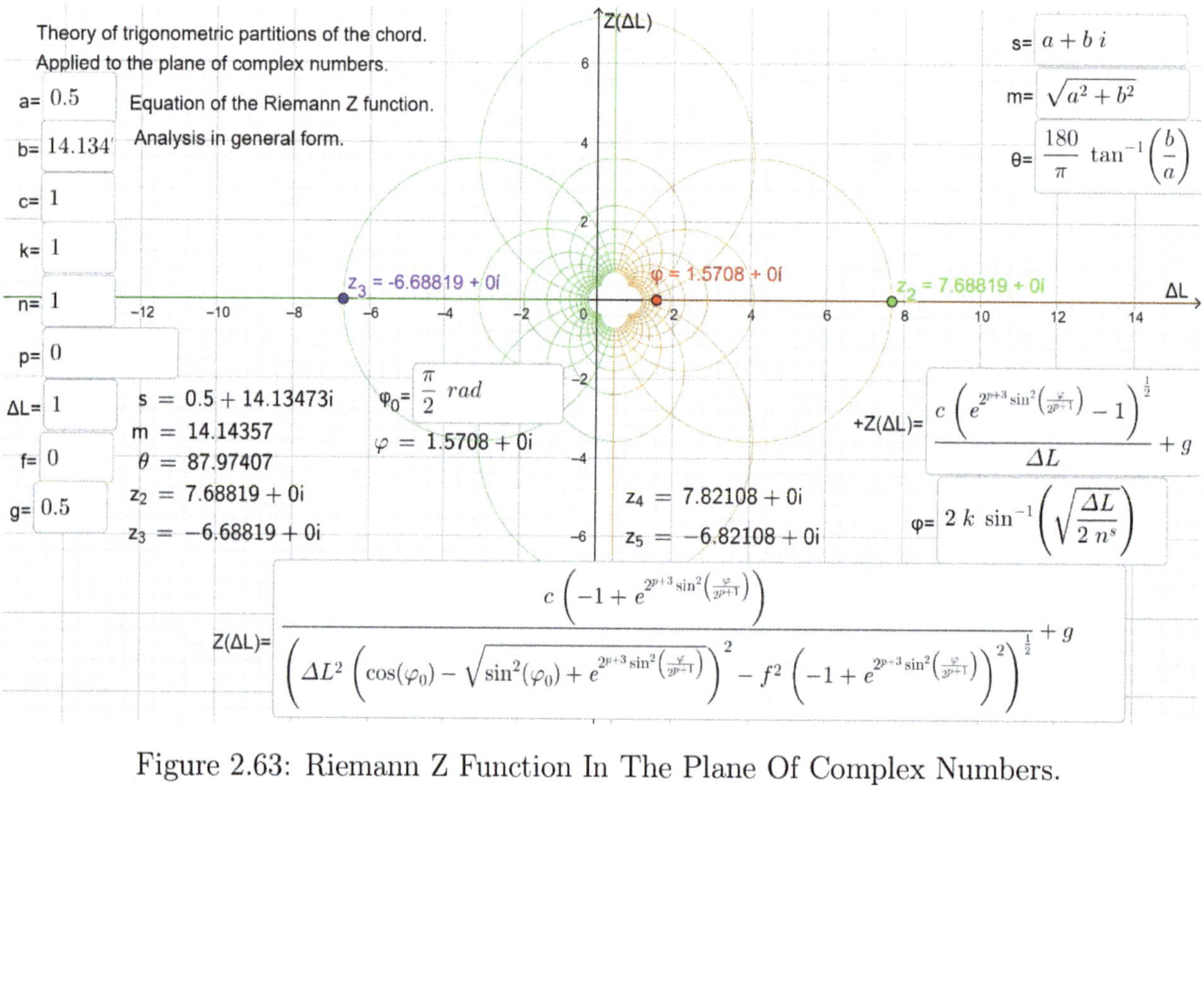

Figure 2.63: Riemann Z Function In The Plane Of Complex Numbers.

With the following prime number values for the variables: n=11, ΔL=4, a=0.5, b=0, s=0.5+0i, p=0, f=0, g=0.5,$\varphi_0 = 0$ or $2 * \pi$ and the rest of the variables equal to one, larger graphs of the Riemann Z function can be generated.

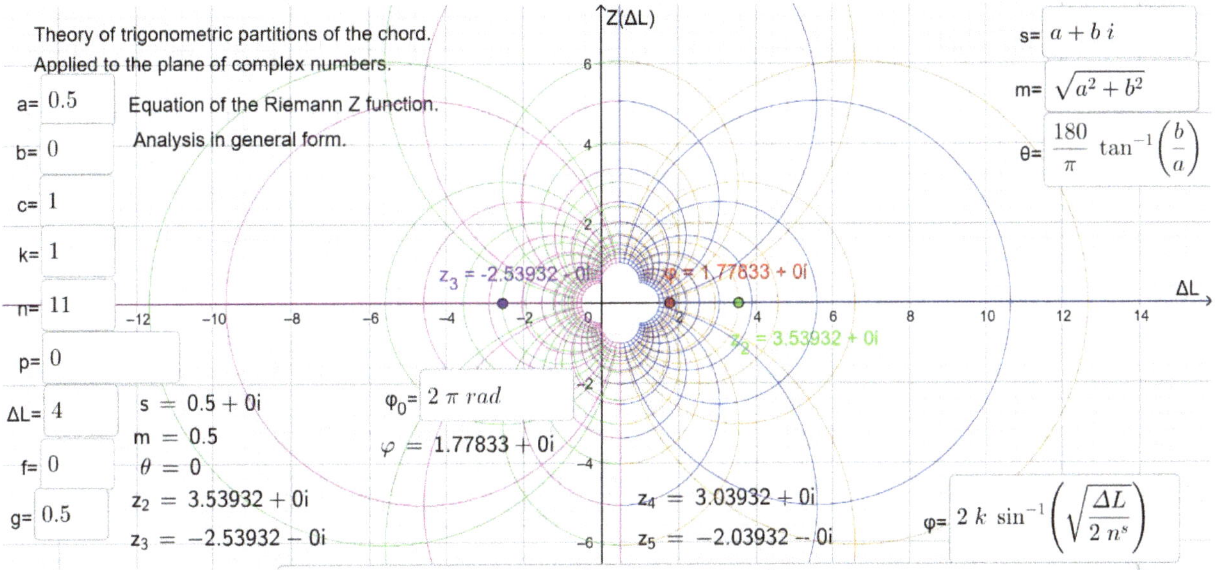

$$Z(\Delta L)=\frac{c\left(-1+e^{2^{p+3}\sin^2\left(\frac{\varphi}{2^{p+1}}\right)}\right)}{\left(\Delta L^2\left(\cos(\varphi_0)-\sqrt{\sin^2(\varphi_0)+e^{2^{p+3}\sin^2\left(\frac{\varphi}{2^{p+1}}\right)}}\right)^2-f^2\left(-1+e^{2^{p+3}\sin^2\left(\frac{\varphi}{2^{p+1}}\right)}\right)^2\right)^{\frac{1}{2}}}+g$$

Figure 2.64: Riemann Z Function In The Plane Of Complex Numbers.

2.3 Analysis Of The Riemann Zeta Function In Relation To Prime Numbers.

The number one, although not considered a prime number, is the base of all prime numbers. In these general equations of the Riemann Z function, the larger the values of the prime numbers used, the smaller the graph generated for the Riemann Z function becomes. The trigonometric partitions **p** do the work of the complex numbers. Additionally, the sign of **p** must be considered based on the direction of rotation of the clock hands, this sign is related to the angular velocity ω, the trigonometric partitions **p**; are those that adjust the precision of the calculations, such as making the height **f** above the **x** axis approach zero or be zero. The variables **n** which is the largest (next) prime number or final radius r and r_1 which is the smallest (previous) prime number or any number, help us to analyze the complete plane of the general equation of the solution of the Riemann function Z. The variables ΔL is the modulus of the length of the prime numbers and is equal to the difference between the largest prime number **n** and the smallest prime number r_1. Also, **n** and r_1 can be any number. The variables **c** and **k** of the general equation of the Riemann Z function are similar in increasing the graph to the desired size. The variable **a** with **b** equal to zero, makes us analyze **s** in the plane of real numbers. The variable **b**, of the general equation of the Riemann Z function, makes the graph rotate.We can combine the variables **b** and **f** to generate new values and verify the types of graphs they produce. The variable **g**, of the general equation of the Riemann Z function, makes the graph move.This constant in the real plane is a horizontal asymptote, and in the complex plane, it is a vertical asymptote, generating additional vertical axes in the graph of the Riemann Z function in the complex plane. The Riemann Z function has a vertical asymptote in the real plane at zero, but in the complex plane, this asymptote is a horizontal asymptote at zero, likewise generating an additional axis on the graph and creating a void at the center of the graph, since the value of $Z(\Delta L)$ is a result at infinity. As we approach this asymptote at zero, the graph of the Riemann Z function starts to grow. The angle φ, is the angle generated from the equation of the growth of the radius as a function of the trigonometric partitions $\Delta L = \Delta r = 2 * r * \sin^2(\varphi/(2^p * k))$ and that relates to the length module of the prime numbers and their rotations **k**. In the prime number length module you can select two adjacent prime numbers or two distant prime numbers, an easy way to generate the graph of the Riemann Z function is considering $r_1 = 0$, **b=0**, $\varphi_0 = 0$, with **n** equal to a prime number. This can be compared with the graph of the Möbius transform or Riemann Sphere. These equations in all their variables can use real and complex numbers. In these equations of the trigonometric partitions for the solution of the Riemann Z function; we can use in GeoGebra the ranges of the variables manually and in automatic or variable. This means two things; the first: is that we can use fixed data or tables that are at the beginning of this book. The second: is that we can use variable data as equations. The tables developed in section 1.2 of the definition of the equations of the prime numbers in the plane of the real numbers, and the trigonometric partitions found in each table are special and help us to generate the graph of the Riemann Z function with the general equations. When we change from the plane of the real numbers to the plane of the imaginary numbers we can change the variable for the functions, for example; if we have the function Z(s) in the real plane which is the invisible graph, or of the points, and if we change from the real plane to the complex plane we can change the variable, for example the variable of the length module of the prime numbers $Z(\Delta L)$; who reveals to us the graph that generates the conjecture or Riemann Z function in relation to prime numbers.

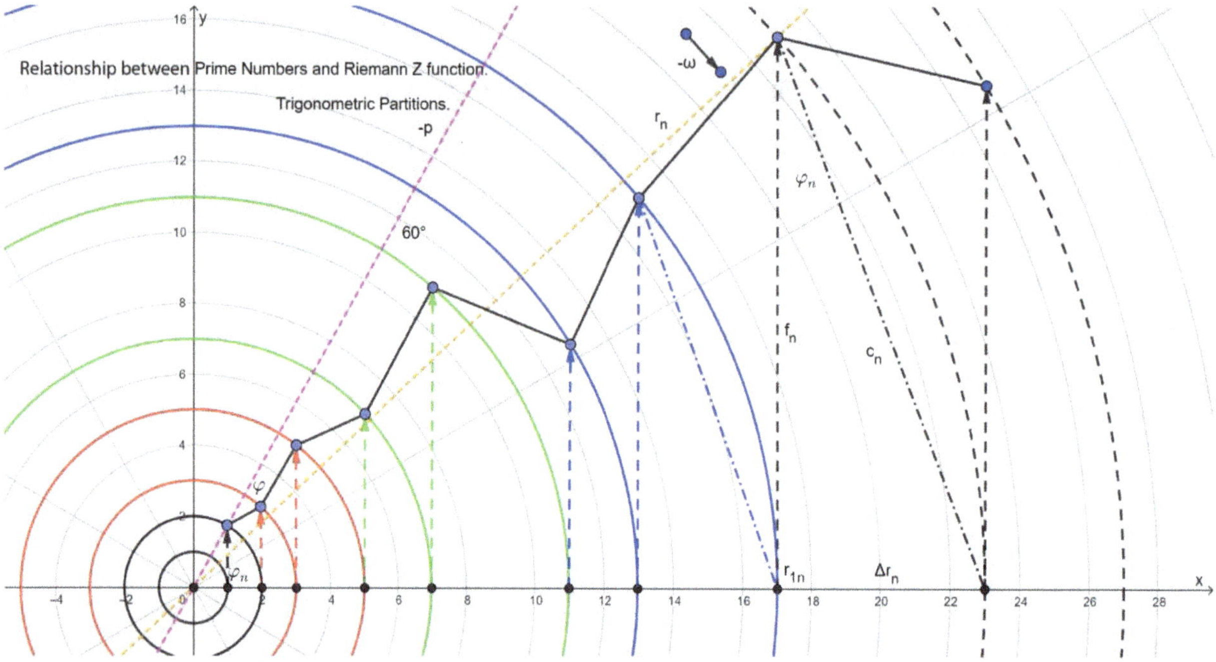

Figure 2.65: Relation Between Prime Numbers and the Riemann Zeta Function. Riemann Zeta Function in the Complex Number Plane.

In this graph of the relationship between prime numbers and the Riemann Z function, we can see the linear graph generated by the union of each point of the prime numbers, we can circularly move these points to any position, such as all the points on the **Y** axis or imaginary axis, and all the points on the **X** axis of the real numbers, where their mathematical analysis is better. The length modulus of the prime numbers supports us to analyze the next prime number minus the previous prime number. Where we can observe the wave pairs with their various generated frequencies. If we analyze with the Theory of Spiral Angles, Spirals, and Trigonometric Partitions each pair of prime numbers, and generate a right triangle, to obtain its nth chord C_n, its nth variable height $Y_n = f_n$, with its nth angle of the trigonometric partitions φ_n, and φ_0 is the initial angle of analysis in the real or complex plane. Therefore, we can say that the nth chord is equal to $c_n = 2 * r_n * \sin(\varphi/2^p)$, Where the initial radius is r_{1n} and the final nth radius is equal to $r_n = r_{1n} * e^{2^{p+2} * \sin^2(\varphi/2^{p+1})}$,$r_n^2/r_{1n}^2 = e^{2^{p+3} * \sin^2(\varphi/2^{p+1})}$, by substituting this equation of the nth radius of the chord, from the theory of trigonometric partitions we can obtain an nth chord in terms of trigonometric partitions of the chord.

Therefore, $c_n = 2 * r_{1n} * \sin(\varphi/2^p) * e^{2^{p+2} * \sin^2(\varphi/2^{p+1})}$. As we require $Y_n = f_n$, applying the Pythagorean theorem, with $f_n = cos(\varphi/2^p) * c_n$ we obtain; that $f_n = r_{1n} * \sin(\varphi/2^p) * e^{2^{p+2} * \sin^2(\varphi/2^{p+1})}$, which is the equation that relates each prime number with its point on the linear graph generated by the prime numbers. This equation of f_n variable is capable of generating scalograms if we write it in terms of the trigonometric partitions $f_n(p)$. This equation of f_n variable can be substituted in the general equations of the Riemann Z function. The objective of the f_n variable is to try to make it zero or almost zero to obtain the values of the trivial zeros, which can be obtained by fixing the smallest prime number (two) and increasing the next prime number, and the non-trivial zeros, which can be obtained by collapsing the mathematical equation, bringing **p** close to its limit with values 1020 or 1022 depending on the equation being used, when we make this change to the trigonometric partitions, the rest of the variables do not cause changes; it remains the same, generating the graph of the Riemann Z function in relation to prime numbers, even if they are not prime numbers.

The variable **f** of $f_n = r_{1n} * \sin(\varphi/2^p) * e^{2^{p+2} * sin^2(\varphi/2^{p+1})}$, with $r_{1n} = \sqrt{f^2 + n^{2s}}$; can be solved, equal to $f = n * \sin(\varphi/2^p)$, if $\Delta L = n - r_{1n} = n - \sqrt{f^2 + n^{2s}}$. However, for the purpose of manual handling and prime number analysis, we will keep it as shown in the graphs of the Riemann Z function. In the polar plane, if we choose the initial angles θ_0 and φ_0, which are the angles formed on the parallel x-axis, and a fixed or constant final angle φ, the trigonometric partitions **p** can be generated along a single axis, following the same direction of increase or decrease of the radius, depending on what is being studied or analyzed (see Figure 2.1). Similarly, if these angles are variable, the trigonometric partitions **p** can be generated radially in the polar plane, dividing the circles into multiple segments. The cutting direction of the partitions depends on whether the angle is constant or variable. Additionally, the correct equation for the final angle φ must be chosen to generate the desired graph, as selecting a different equation may result in an undesired graph.

Negative quadratic solution: The graph in the complex plane will be composed of two halves, one positive $+Z(\Delta L)$ and another negative $-Z(\Delta L)$. With the values: $a = 1/2$, $b = 0$ which produces rotation of the graph, with s=1/2+0i, $c = 29$ to increase the size of the graph, with the previous prime $r_1 = 2$, with the subsequent prime n=3, with f_n variable by the height equation Y_n, as a function of the trigonometric partitions, with the length module $\Delta L = 1$. With the trigonometric partitions in a clockwise direction p=-1, with the graph shifted with g=0.5 on the x-axis, with the initial analysis of the angle $\varphi_0 = \pi/2$. The nth angle φ; is generated by the equation of the trigonometric partitions of the growth of the radius.Equation of the trigonometric partitions of the chord:

$$Z(\Delta L) = \sum_{n=1}^{\infty} (1/n^s) = c * [e^{2(p+3)*\sin^2(\varphi/2^{p+1})} - 1]$$
$$/\sqrt{\Delta L^2 * [\cos(\varphi_0) - \sqrt{[\sin^2(\varphi_0) + e^{2(p+3)*\sin^2(\varphi/2^{p+1})}]}]^2 - f^2 * [e^{2(p+3)*\sin^2(\varphi/2^{p+1})} - 1]^2} + g.$$

Solution of the Negative Square Root: Left side of the graph.

$$Z(\Delta L) = \sum_{n=1}^{\infty} (1/n^s) = -c * [e^{2(p+3)*\sin^2(\varphi/2^{p+1})} - 1]$$
$$/\sqrt{\Delta L^2 * [\cos(\varphi_0) - \sqrt{[\sin^2(\varphi_0) + e^{2(p+3)*\sin^2(\varphi/2^{p+1})}]}]^2 - f^2 * [e^{2(p+3)*\sin^2(\varphi/2^{p+1})} - 1]^2} + g.$$

When we assign numerical values to each variable in the general equation of the Riemann Z function, it is equivalent to searching for coefficients to find various mathematical equations in order to generate the graph of the Riemann Z function.

If we are in the complex number plane and substitute these mathematically equivalent angle equations, it leads us to the Cartesian plane of real numbers. With this equation, in some cases, we can generate graphs of hyperbolic functions. If we use these equivalent angle equations in the equation of Figure 2.66, a graph can be generated with vertical asymptotes at $1/2$ and $-1/2$, and a horizontal asymptote at zero.That is the reason why, in these formulas, the graphs are generated in two halves due to their two asymptotes, as seen in this graph. With $\varphi = 2\pi n = \omega t = 2\pi f t = kx - \omega t$, where $k = \frac{2\pi}{\lambda}$, we can perform an analysis in the Cartesian plane of real numbers as a function of angular velocity and time, frequency and time, position and time, the wave number k, and λ, its wavelength.

Every trivial and non-trivial zero can be obtained, reviewed, and verified in the GeoGebra program, where each formula was developed.

These equations prove that the Riemann Hypothesis is true.

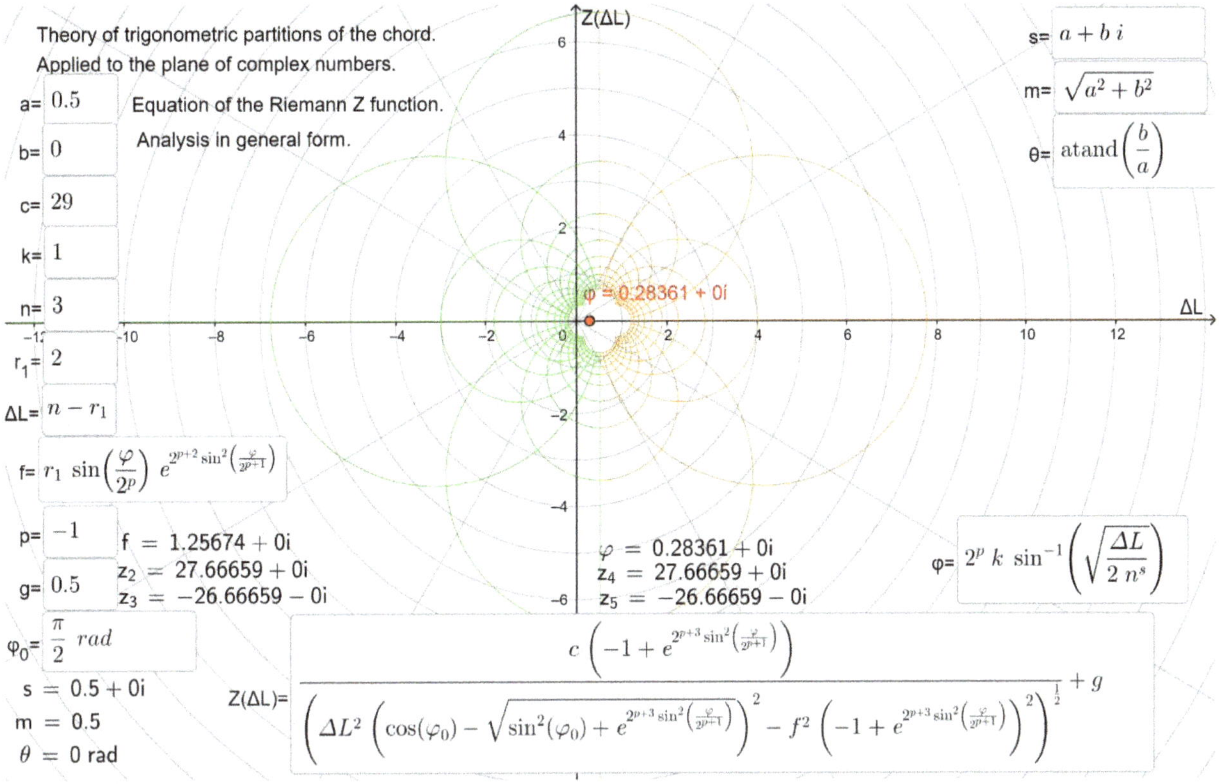

Figure 2.66: Relation Between Prime Numbers and the Riemann Zeta Function. Analysis in the Complex Number Plane for Negative Equations.

Positive quadratic solution: The graph in the complex plane will be composed of two halves, one positive $+Z(\Delta L)$ and one negative $-Z(\Delta L)$. With the values: $a = 1/2$, $b = 0$ which rotates the graph, with s=1/2+0i, $c = 29$ to increase the size of the graph, with the previous prime $r_1 = 2$, with the subsequent prime n=3, with f_n variable by the height equation Y_n, as a function of the trigonometric partitions, with the length module $\Delta L = 1$. With the trigonometric partitions clockwise p=-1, with the graph moved with g=0.5 on the x-axis, with the initial analysis of the angle $\varphi_0 = \pi/2$. The nth angle φ; is generated by the equation of the trigonometric partitions of the growth of the radius.

Equation of the trigonometric partitions of the chord:

$$Z(\Delta L) = \sum_{n=1}^{\infty} (1/n^s) = c * [e^{2(p+3)*\sin^2(\varphi/2^{p+1})} - 1]$$
$$/\sqrt{\Delta L^2 * [\cos(\varphi_0) + \sqrt{[\sin^2(\varphi_0) + e^{2(p+3)*\sin^2(\varphi/2^{p+1})}]}]^2 - f^2 * [e^{2(p+3)*\sin^2(\varphi/2^{p+1})} - 1]^2} + g.$$

Solution of the Positive Square Root: Left side of the graph.

$$Z(\Delta L) = \sum_{n=1}^{\infty} (1/n^s) = -c * [e^{2(p+3)*\sin^2(\varphi/2^{p+1})} - 1]$$
$$/\sqrt{\Delta L^2 * [\cos(\varphi_0) + \sqrt{[\sin^2(\varphi_0) + e^{2(p+3)*\sin^2(\varphi/2^{p+1})}]}]^2 - f^2 * [e^{2(p+3)*\sin^2(\varphi/2^{p+1})} - 1]^2} + g.$$

Analytical Geometry now has the equations of Spiral Angles, Spirals, and Trigonometric Partitions to analyze the entire plane of complex numbers of the Riemann conjecture and other mathematical phenomena. In the equation Z(ΔL) which is the function of the modulus of length of the prime numbers for the Riemann Z function or $\zeta(\Delta P)$; in all its variables, if the largest amount of prime numbers is used, the graph that will be created will be that of the Riemann Z Function, which is related to the Riemann sphere and Möbius transformation. When selecting the signs of the numbers to be used for each variable in these equations, great care must be taken to assign the correct sign in order to create the graph of the Riemann Z function. An additional equation is the **Y** component of the equation for the trigonometric partitions of the radius growth, allowing for a variable analysis of $Y_n = f_n$ and verifying the consistency of the solution of the general equation of the Riemann Z function. These same equations for the **Y** component as a function of the angle generate wavelets, and as a function of the trigonometric partitions, they generate scalograms.

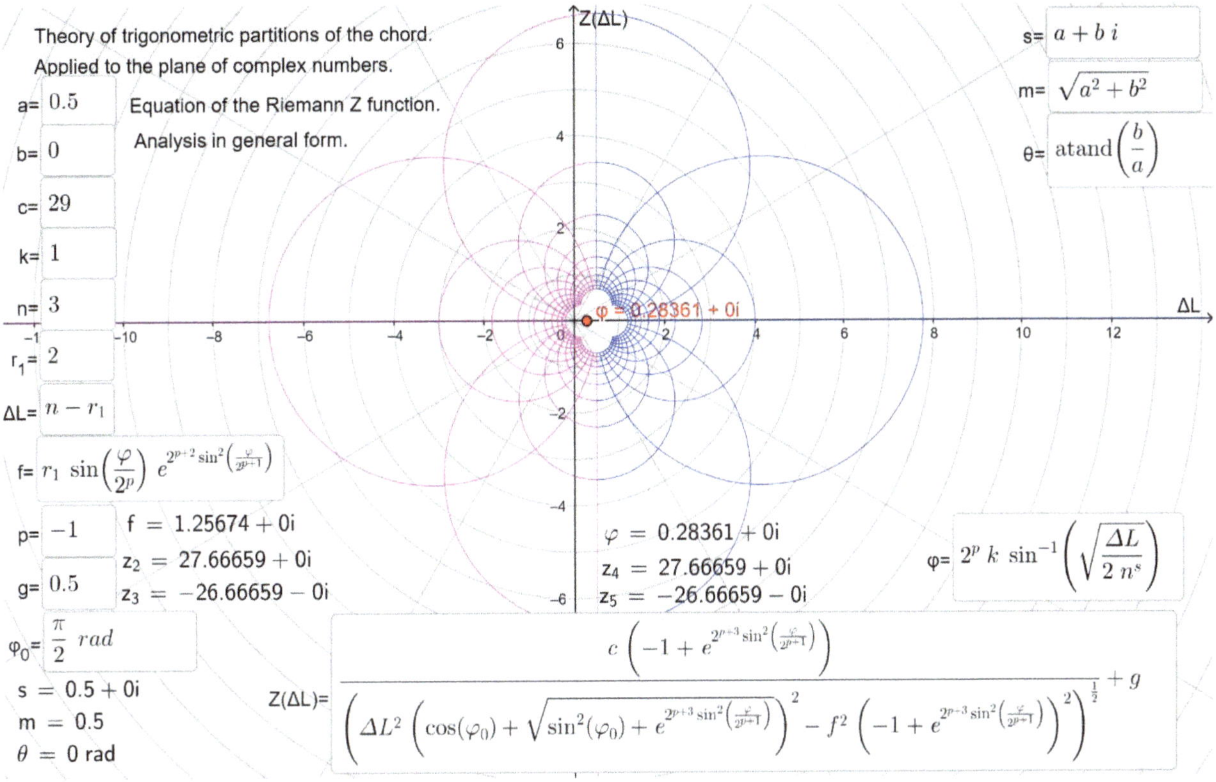

Figure 2.67: Relation Between Prime Numbers and the Riemann Zeta Function. Analysis in the Complex Number Plane for Positive Equations.

It is observed that the larger the prime number used in these equations of the general solution of the Riemann Z function, the smaller the graph generated by the Riemann Z function. Its size can also be controlled with the variable **c**.

By following the same process carried out in this book and using the equations of trigonometric partitions including those related to the circular sector, the chord, the apothem, the radial growth, the arrow, and other equations related to radii, such as the apothem plus the arrow in relation to trigonometric partitions we can generate the graph that represents the Riemann sphere, the Möbius transformation, or the function $F(Z) = \frac{1}{Z}$ in the complex number plane.

The equation for the radius growth in relation to the trigonometric partitions is:

$r = r_1 * e^{\varphi - 2^{(p-1)} * \sin(\varphi/2^{(p-1)})}, r^2/r_1^2 = e^{2*\varphi - 2^{(p)} * \sin(\varphi/2^{(p-1)})}$

The **X** component: $r_x = r_1 * \cos(\varphi/2^p) * e^{\varphi - 2^{(p-1)} * \sin(\varphi/2^{(p-1)})}$

The **Y** component: $Y_n = f_n = r_y = r_1 * \sin(\varphi/2^p) * e^{\varphi - 2^{(p-1)} * \sin(\varphi/2^{(p-1)})}$.

Examples: Equation of the trigonometric partitions of radius growth: Positive solution of the quadratic equation:

$$Z(\Delta L) = \sum_{n=1}^{\infty}(1/n^s) = c * [e^{2*\varphi - 2^p * \sin(\varphi/2^{(p-1)})} - 1] / \sqrt{\Delta L^2 * [\cos(\varphi_0) + \sqrt{[\sin^2(\varphi_0) + e^{2*\varphi - 2^p * \sin(\varphi/2^{(p-1)})}]}]^2 - f^2 * [e^{2*\varphi - 2^p * \sin(\varphi/2^{(p-1)})} - 1]^2 + g}.$$

Positive square root solution: Left side of the graph.

$$Z(\Delta L) = \sum_{n=1}^{\infty}(1/n^s) = -c * [e^{2*\varphi - 2^p * \sin(\varphi/2^{(p-1)})} - 1] / \sqrt{\Delta L^2 * [\cos(\varphi_0) + \sqrt{[\sin^2(\varphi_0) + e^{2*\varphi - 2^p * \sin(\varphi/2^{(p-1)})}]}]^2 - f^2 * [e^{2*\varphi - 2^p * \sin(\varphi/2^{(p-1)})} - 1]^2 + g}.$$

Negative solution of the quadratic equation:

$$Z(\Delta L) = \sum_{n=1}^{\infty}(1/n^s) = c * [e^{2*\varphi - 2^p * \sin(\varphi/2^{(p-1)})} - 1] / \sqrt{\Delta L^2 * [\cos(\varphi_0) - \sqrt{[\sin^2(\varphi_0) + e^{2*\varphi - 2^p * \sin(\varphi/2^{(p-1)})}]}]^2 - f^2 * [e^{2*\varphi - 2^p * \sin(\varphi/2^{(p-1)})} - 1]^2 + g}.$$

Positive square root solution: Left side of the graph.

$$Z(\Delta L) = \sum_{n=1}^{\infty}(1/n^s) = -c * [e^{2*\varphi - 2^p * \sin(\varphi/2^{(p-1)})} - 1] / \sqrt{\Delta L^2 * [\cos(\varphi_0) - \sqrt{[\sin^2(\varphi_0) + e^{2*\varphi - 2^p * \sin(\varphi/2^{(p-1)})}]}]^2 - f^2 * [e^{2*\varphi - 2^p * \sin(\varphi/2^{(p-1)})} - 1]^2 + g}.$$

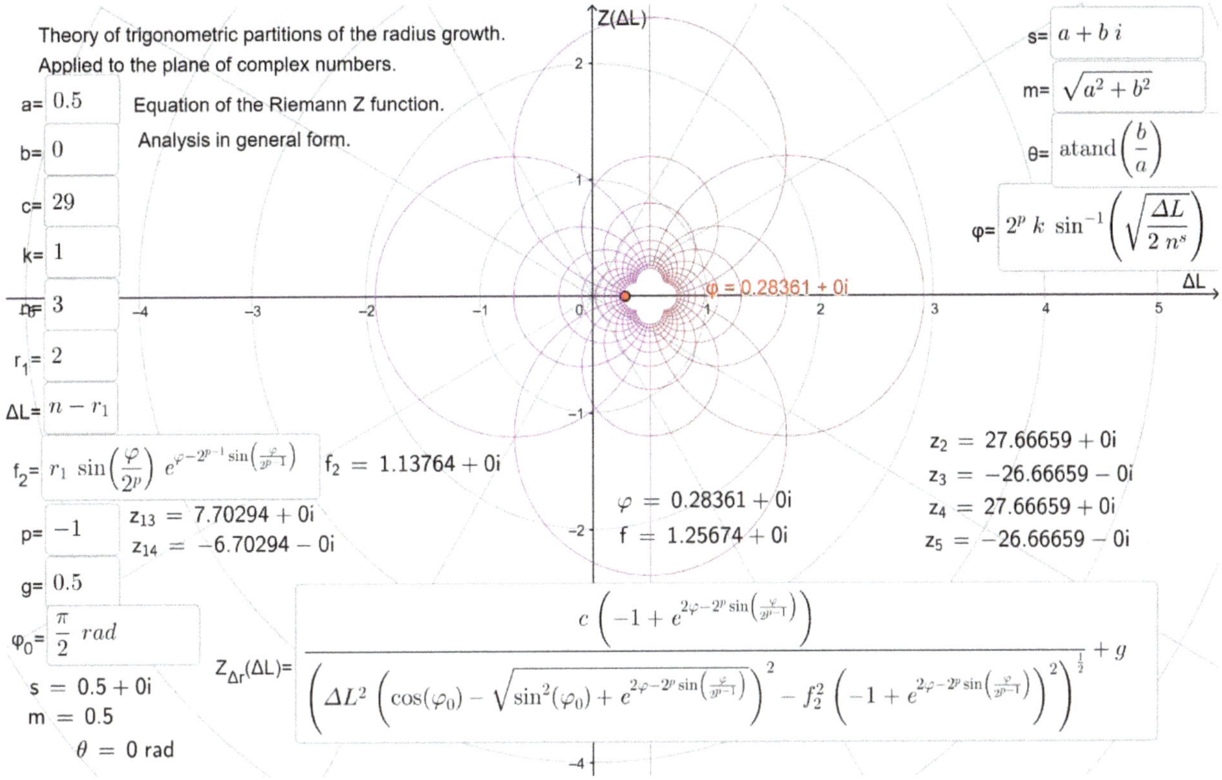

Figure 2.68: Relation Between Prime Numbers and the Riemann Zeta Function. Analysis in the Complex Number Plane for Negative Equations.

Similarly, we can apply the same approach to the equation of the arrow partitions: Each of the equations of trigonometric partitions of the circular sector, the chord, the apothem, and the arrow helps us study each prime number differently and analyze the effects that each prime or non-prime number has on the graph of the Riemann Zeta function. With the equations of the trigonometric partitions of the arrow, it is observed that as we move from a small prime number to a larger one, the graph stops rotating and becomes more stable along the **x** and **y** axes. The radius equation for the arrow, in relation to trigonometric partitions, is:

$r = r_1 * e^{\varphi - 2^p * \sin(\varphi/2^p)}, r^2/r_1^2 = e^{2*\varphi - 2^{(p+1)} * \sin(\varphi/2^p)}$

The **X** component:

$r_x = r_1 * \cos(\varphi/2^p) * e^{\varphi - 2^p * \sin(\varphi/2^p)}$

The **Y** component:

$$Y_n = f_n = r_y = r_1 * \sin(\varphi/2^p) * e^{\varphi - 2^P * \sin(\varphi/2^P)}.$$

Examples: Positive solution of the quadratic equation:

$$Z(\Delta L) = \textstyle\sum_{n=1}^{\infty}(1/n^s) = c * [e^{2*\varphi - 2^{(p+1)} * \sin(\varphi/2^p)} - 1]$$

$$/\sqrt{\Delta L^2 * [\cos(\varphi_0) + \sqrt{[\sin^2(\varphi_0) + e^{2*\varphi - 2^{(p+1)} * \sin(\varphi/2^p)}]]}^{\,2} - f^2 * [e^{2*\varphi - 2^{(p+1)} * \sin(\varphi/2^p)} - 1]^2 + g}.$$

Positive square root solution: Left side of the graph.

$$Z(\Delta L) = \textstyle\sum_{n=1}^{\infty}(1/n^s) = -c * [e^{2*\varphi - 2^{(p+1)} * \sin(\varphi/2^p)} - 1]$$

$$/\sqrt{\Delta L^2 * [\cos(\varphi_0) + \sqrt{[\sin^2(\varphi_0) + e^{2*\varphi - 2^{(p+1)} * \sin(\varphi/2^p)}]]}^{\,2} - f^2 * [e^{2*\varphi - 2^{(p+1)} * \sin(\varphi/2^p)} - 1]^2 + g}.$$

Negative solution of the quadratic equation:

$$Z(\Delta L) = \textstyle\sum_{n=1}^{\infty}(1/n^s) = c * [e^{2*\varphi - 2^{(p+1)} * \sin(\varphi/2^p)} - 1]$$

$$/\sqrt{\Delta L^2 * [\cos(\varphi_0) - \sqrt{[\sin^2(\varphi_0) + e^{2*\varphi - 2^{(p+1)} * \sin(\varphi/2^p)}]]}^{\,2} - f^2 * [e^{2*\varphi - 2^{(p+1)} * \sin(\varphi/2^p)} - 1]^2 + g}.$$

Negative square root solution: Left side of the graph.

$$Z(\Delta L) = \textstyle\sum_{n=1}^{\infty}(1/n^s) = -c * [e^{2*\varphi - 2^{(p+1)} * \sin(\varphi/2^P} - 1]$$

$$/\sqrt{\Delta L^2 * [\cos(\varphi_0) - \sqrt{[\sin^2(\varphi_0) + e^{2*\varphi - 2^{(p+1)} * \sin(\varphi/2^p)}]]}^{\,2} - f^2 * [e^{2*\varphi - 2^{(p+1)} * \sin(\varphi/2^p)} - 1]^2 + g}.$$

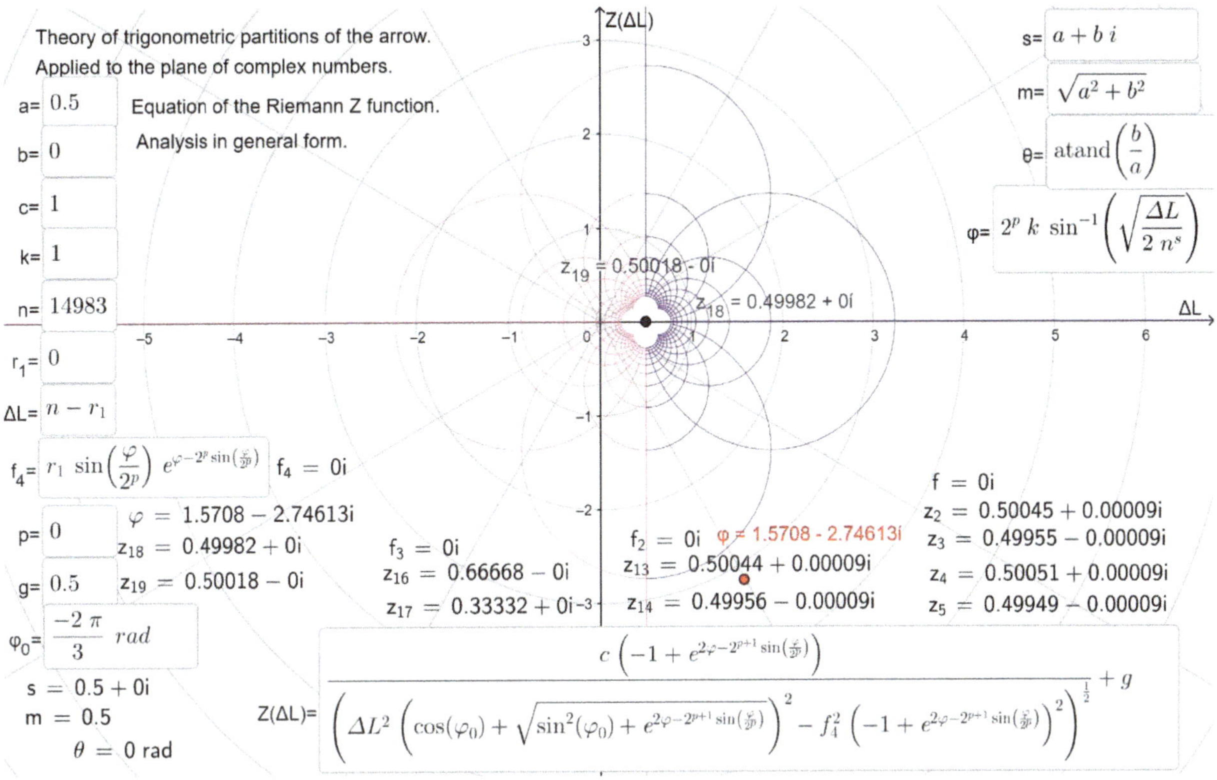

Figure 2.69: Relation Between Prime Numbers and the Riemann Zeta Function. Analysis in the Complex Number Plane for Positive Equations.

Similarly, we can apply the same approach to the equation of the apothem partitions: The equation of the apothem in relation to the radius. The chord divides the radius into two parts, creating the Apothem and the Chord. The apothem originates at the center with $r_1 = 0$, extending to any value within the space of the circle with its corresponding angle. This allows us to analyze numbers, whether prime or non-prime, in the form of circular waves. The equations of the trigonometric partitions of the apothem are among the most precise for analyzing the Riemann Zeta function in a circular form, where each number in the Cartesian Plane can be real or complex. This analysis considers each number from the center to the final length of the radius.One of the characteristics of this equation, related to the trigonometric partitions of the apothem, is that the more cuts **p** are made, the angle tends to zero. Therefore, at infinity, the apothem tends to become the radius, which is why the graph of the Riemann Z function tends to grow. The opposite happens with the rest of the components of a circle, such as the circular sector, the chord, the arrow, and the growth of the radius; as more cuts or partitions are made to the circle, all these components tend to zero at infinity. This is why the graphs of the Riemann Z function become smaller until they shrink to a point or disappear.

$r = r_1 * e^{2P*\sin(\varphi/2^p)}, r^2/r_1^2 = e^{2(p+1)*\sin(\varphi/2^P)}$

The **X** component:

$r_x = r_1 * \cos(\varphi/2^p) * e^{2^P*\sin(\varphi/2^p)}$

The **Y** component:

$Y_n = f_n = r_y = r_1 * \sin(\varphi/2^p) * e^{2^P*\sin(\varphi/2^p)}$.

The apothem in relation to the trigonometric partitions is: Examples: Positive solution of the quadratic equation: $Z(\Delta L) = \sum_{n=1}^{\infty}(1/n^s) = c * [e^{2(p+1)*\sin(\varphi/2^p)} - 1]$

$/\sqrt{\Delta L^2 * [\cos(\varphi_0) + \sqrt{[\sin^2(\varphi_0) + e^{2(p+1)*\sin(\varphi/2^p)}]]}^2 - f^2 * [e^{2(p+1)*\sin(\varphi/2^p)} - 1]^2 + g}$.

Positive square root solution: Left side of the graph.

$Z(\Delta L) = \sum_{n=1}^{\infty}(1/n^s) = -c * [e^{2(p+1)*\sin(\varphi/2^p)} - 1]$

$/\sqrt{\Delta L^2 * [\cos(\varphi_0) + \sqrt{[\sin^2(\varphi_0) + e^{2(p+1)*\sin(\varphi/2^p)}]]}^2 - f^2 * [e^{2(p+1)*\sin(\varphi/2^p)} - 1]^2 + g}$.

Negative solution of the quadratic equation:

$Z(\Delta L) = \sum_{n=1}^{\infty}(1/n^s) = c * [e^{2(p+1)*\sin(\varphi/2^p)} - 1]$

$/\sqrt{\Delta L^2 * [\cos(\varphi_0) - \sqrt{[\sin^2(\varphi_0) + e^{2(p+1)*\sin(\varphi/2^p)}]]}^2 - f^2 * [e^{2(p+1)*\sin(\varphi/2^p)} - 1]^2 + g}$.

Negative square root solution: Left side of the graph.

$Z(\Delta L) = \sum_{n=1}^{\infty}(1/n^s) = -c * [e^{2(p+1)*\sin(\varphi/2^p} - 1]$

$/\sqrt{\Delta L^2 * [\cos(\varphi_0) - \sqrt{[\sin^2(\varphi_0) + e^{2(p+1)*\sin(\varphi/2^p)}]]}^2 - f^2 * [e^{2(p+1)*\sin(\varphi/2^p)} - 1]^2 + g}$.

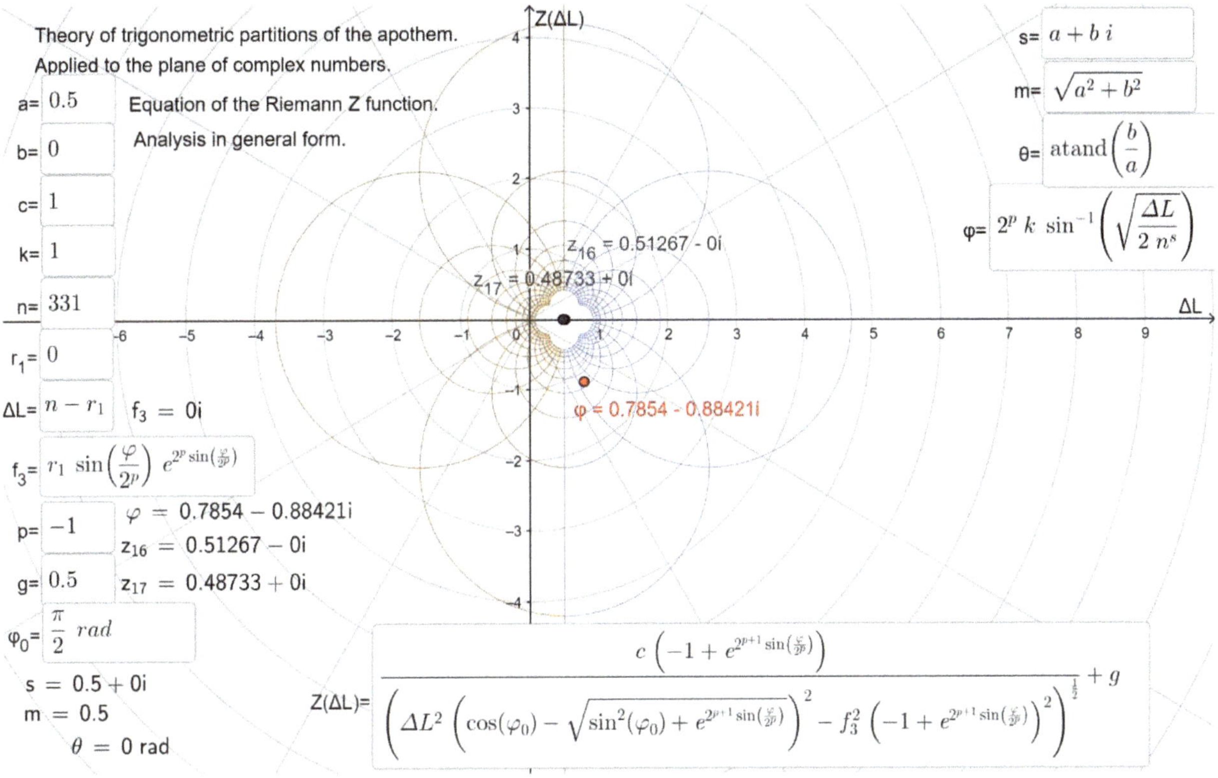

Figure 2.70: Relation Between Prime Numbers and the Riemann Zeta Function. Analysis in the Complex Number Plane for Negative Equations.

And likewise, with the rest of the discovered trigonometric partition equations, such as the one where the radius equals the apothem plus the chord, we can continue generating the graph of the Riemann Zeta function.

Finally, the Theory of Spiral Angles, Spirals, and Trigonometric Partitions, in its equations concerning the circular sector, the chord, the apothem, the arrow, and the growth of the radius... regarding the solution to the Riemann Z function, shows us that we can perform a transformation related to the squared power of the ratio between the largest prime number n and the smallest prime number r_1 to study prime numbers. It is worth mentioning that this squared ratio can also be either prime or non-prime, which is relevant for studying number theory. In these transformations of the squared ratio of prime numbers, we can express this solution as a function of the length modulus $Z(\Delta L)$, which generates graphs similar to those produced by the equations of trigonometric partitions. Therefore, the graphs generated by the equations of trigonometric partitions are identical to those generated by the Riemann sphere and the Z function of Riemann or the function F(z)=1/Z in the complex plane. Moreover, if we define the function of the prime number ratio in terms of s, as Z(s), we observe that it generates graphs rather than discrete points in the complex number plane. These graphs, however, are not identical to those of the Riemann Z function. Substituting this ratio into the equation of the trigonometric partitions in the general form analysis of the Riemann Z function, the following equation is obtained:

$r^2/r_1^2 = e^{2^{(p+1)}*\sin(\varphi/2^p)} \approx e^{2*\varphi - 2^{(p+1)}*\sin(\varphi/2^p)} \ldots \approx n^2/r_1^2$.Radial form of $f^2 = n^2 * \sin^2(\varphi/2^p)$

To obtain the best results in the calculations of the non-trivial zeros, the trigonometric partitions **p** should be chosen so that the angle φ approaches zero along the x-axis or is equal to zero.

Examples:

Positive solution of the quadratic equation::

$$Z(\Delta L) = \sum_{n=1}^{\infty}(1/n^s) = c * [(n^2/r_1^2)^2 - 1]$$
$$/\sqrt{\Delta L^2 * [\cos(\varphi_0) + \sqrt{[\sin^2(\varphi_0) + (n^2/r_1^2)^2]}]^2 - n^2 * \sin^2(\varphi/2^p) * [(n^2/r_1^2)^2 - 1]^2 + g}.$$

Positive square root solution: Left side of the graph.

$$Z(\Delta L) = \sum_{n=1}^{\infty}(1/n^s) = -c * [(n^2/r_1^2)^2 - 1]$$
$$/\sqrt{\Delta L^2 * [\cos(\varphi_0) + \sqrt{[\sin^2(\varphi_0) + (n^2/r_1^2)^2]}]^2 - n^2 * \sin^2(\varphi/2^p)) * [(n^2/r_1^2)^2 - 1]^2 + g}.$$

Negative solution of the quadratic equation.

$$Z(\Delta L) = \sum_{n=1}^{\infty}(1/n^s) = c * [(n^2/r_1^2)^2 - 1]$$
$$/\sqrt{\Delta L^2 * [\cos(\varphi_0) - \sqrt{[\sin^2(\varphi_0) + (n^2/r_1^2)^2]}]^2 - n^2 * \sin^2(\varphi/2^p) * [(n^2/r_1^2)^2 - 1]^2 + g}.$$

Negative square root solution: Left side of the graph.

$$Z(\Delta L) = \sum_{n=1}^{\infty}(1/n^s) = -c * [(n^2/r_1^2)^2 - 1]$$
$$/\sqrt{\Delta L^2 * [\cos(\varphi_0) - \sqrt{[\sin^2(\varphi_0) + (n^2/r_1^2)^2]}]^2 - n^2 * \sin^2(\varphi/2^p) * [(n^2/r_1^2)^2 - 1]^2 + g}.$$

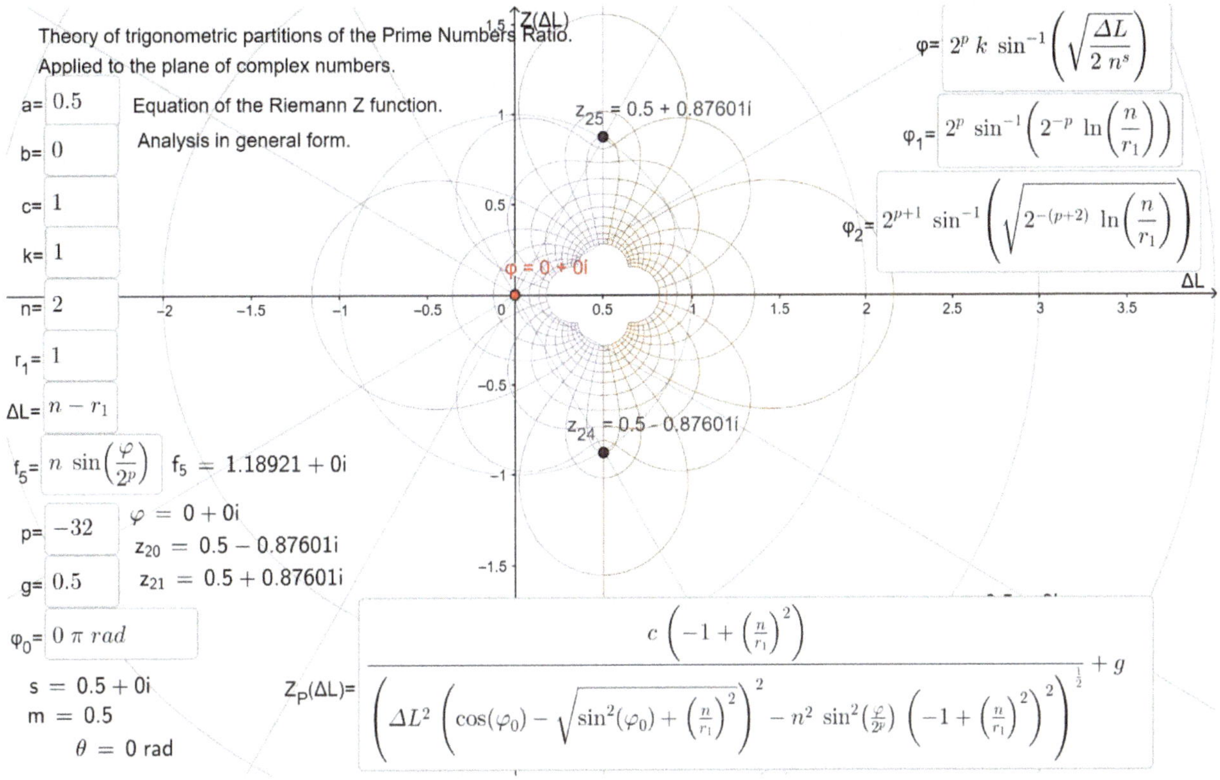

Figure 2.71: Relation Between Prime Numbers and the Riemann Zeta Function. Analysis in the Complex Number Plane for Negative Equations.

The elaborate technique of the general solution, using the Theory of Spiral Angles, Spirals, and Trigonometric Partitions of the Riemann Z function, shows us that various equations can be used in its ratio. For example, if we have the equation of a spiral as $r = r_1 * e^{\varphi^2/2}$, and we follow the same process of squaring the ratios and obtaining their ratios to substitute them into the general equation of the hypothesis solution,$r^2/r_1^2 = e^{\varphi^2}$, we can still obtain graphs similar to those of the Riemann Z function in the complex plane with $Z(\Delta L)$. The graphs created with this same spiral equation in the Z(s) plane almost always show a pattern similar to that of the prime number ratio. The general equation of trigonometric partitions as a function of the prime number ratio can be rewritten in a simplified form by setting $\varphi_0 = 0$, $\varphi = 0$, which is equivalent to bringing all function values to the x-axis. Then, f = 0, with g=0.5, simplifying the formula to $Z(\Delta L) = c * \frac{\sqrt{[(\frac{n}{r_1})^2 - 1]}}{\Delta L} + g$, or $Z(\Delta P) = c * \frac{\sqrt{[(\frac{P}{P_0})^2 - 1]}}{P - P_0} + g$, Where P represents the prime numbers,Which is a hyperbolic function in the plane of real numbers with horizontal and vertical asymptotes at 1/2, 0. and making the function Z(s) disappear. If we take the limit to infinity of the prime numbers, we find that this part of the equation $\frac{\sqrt{[(\frac{n}{r_1})^2-1]}}{\Delta L}$ tends to zero, thus demonstrating that the results of the function $Z(\Delta L)$ will always be $\frac{1}{2}$. Obtaining the limit at infinity of the prime numbers, the function will always tend to the value of the variable g.With this equation and the primes two and three, we can obtain the golden number and its reciprocal. In the GeoGebra program, if we set all variable domains to automatic except for c, after a certain time, the graph turns into a point and later disappears, as the equation takes imaginary numbers. These numbers change the dimension of the function, and we observe that the variable **f** represents the previous prime number, causing large integer numbers to be generated. This indicates that it is important to properly adjust the function's domains to generate a better analysis of prime and non-prime numbers in the equations of the solution of the Riemann Z function.

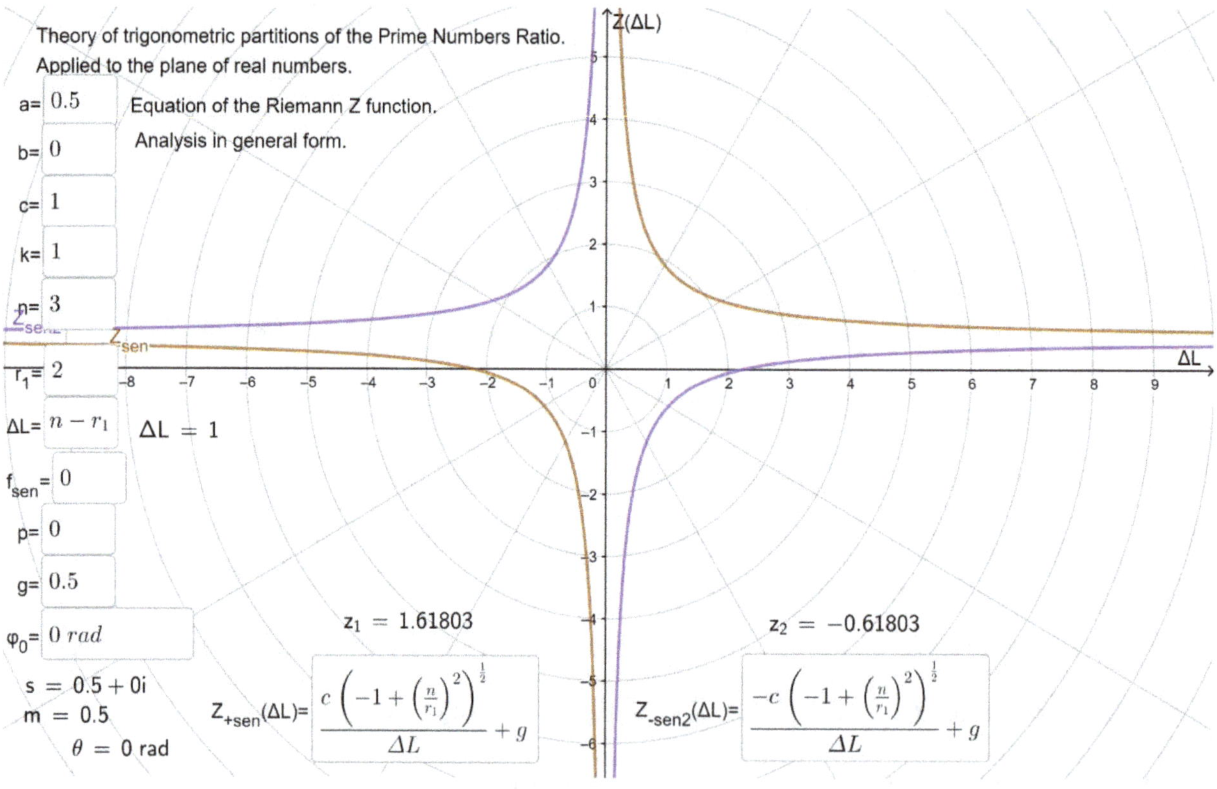

Figure 2.72: Hyperbolic Graph of Prime Numbers. Positive and Negative Equations of the Z Function of Riemann for the Ratio of Prime Numbers.

When we maximize the trigonometric partitions and the initial analysis angle φ_0, which is equal to zero, the values of the final angle φ become extremely high or infinite, as does its frequency. I have called these angles: spiral angles. With these values of the final angles, the partitions, the initial angle, the value of **s**, and the constant **g**, we can calculate all the values of the trivial and non-trivial zeros of the Riemann Z function. There are two ways to calculate the trivial and non-trivial values of the Riemann Z function:

1. Using only prime numbers, with the equations of the positive and negative solutions of the trigonometric partitions of the ratio of prime numbers. This method employs an equation that operates solely in the real plane and is independent of the angle φ. It only requires the ratios of greater or lesser prime numbers and their length modulus, along with the constant **g**, which represents an asymptote in the real or imaginary plane. With this equation, the graph of the Riemann Z function is generated in a single part, using the equations of the trigonometric partitions of the chord, the apothem, the arrow, the radial growth, etc.
2. Using only prime numbers, with the equations of the positive and negative solutions of the trigonometric partitions of the ratio of prime numbers. In these graphs, we can observe how the graph tends to the horizontal asymptote at g=0.5 and the vertical asymptote at 0. The central lines that form in the graph of the Riemann Z function are due to these horizontal and vertical asymptotes. For this reason, the graph can be generated in two parts or halves: the positive part on the right side tends to the value of **g** from its right side, and the negative part on the left side tends to the value of **g** from its left side. Likewise, if we use prime numbers as negative values, we will observe that the equation shifts from the real plane to the imaginary plane. This solution to the Riemann Z function allows us to analyze both the real and imaginary planes. If we notice, the sum of the Z values generated by the equations for the right and left sides is always equal to 1.

(Note: For this, I have developed an Excel spreadsheet.)

If we continue analyzing the equation obtained from the Theory of Trigonometric Partitions and change the prime numbers in a radial form, using the equations of the apothem and the chord, which allow us to solve for the angle φ, we find that when r=n is the larger prime and $r_n = n$ is the smaller prime, we obtain two new equations for the angle φ. These equations can be used in the general equations to analyze prime numbers. We will observe that with these equations, we can generate the same graph as the one produced by the Riemann Zeta function. This means that there is not just one equation that can generate the same pattern as the graph of the Riemann Zeta function. The equation of the apothem is:

$r_n = r_{1n} * e^{2P*\sin(\varphi/2^P)}$

$n = r_1 * e^{2P*\sin(\varphi/2^P)}$

Solving for the angle φ, we obtain:

$\varphi_1 = 2^p * \arcsin[2^{-p} * \ln(n/r_1)]$

The equation of the chord is:

$r_n = r_{1n} * e^{2^{p+2}*\sin^2(\varphi/2^{p+1})}$

$n = r_1 * e^{2^{p+2}*\sin^2(\varphi/2^{p+1})}$

Solving for the angle φ, we obtain:

$\varphi_2 = 2^{(p+1)} * \arcsin\sqrt{[2^{-(p+2)} * \ln(n/r_1)]}$

By substituting these equations for the angles and analyzing each within the equation of trigonometric partitions of the prime number ratio, we can generate a different pattern from the graph of the Riemann Zeta function. In the book Spiral Angles, Spirals, and Trigonometric Partitions, I explain the importance of using the correct equation for the spiral angle, as different equations can yield different graphs. The partitions p are a variable that makes corrections in numerical calculations and provides the precision needed to obtain the correct value. This same process must be applied to the positive equations of the trigonometric partitions of the prime number ratio to obtain the graphs of the patterns of the Riemann Zeta function.

If we change this function $Z(\Delta L)$ from the trigonometric partitions of the ratio of prime numbers as a function of s; $Z(s)$, and use the angle φ_2, the pattern of the graph of Riemann's Z function transforms into two infinite horizontal lines at the values zero and one. When we change variables in the functions ΔL and s, we are transitioning from the imaginary plane to the real plane or vice versa; that is, from the imaginary dimension to the real dimension or vice versa.

In the Cartesian plane, if we set $x = n^s$ and $y = f$, applying the Pythagorean theorem, we obtain the tangential form $f = n^s \tan(\varphi/2^p)$, with the trigonometric partitions **p** at their maximum, where $p = 1022$. The final angles φ will be extremely large and infinite, or likewise, may tend to zero, depending on the equation used. The same applies to their frequencies. In this way, we make **f** tend to zero, and in the equation $s = a + bi$, **b** can take any value without affecting the graph of the Riemann Zeta function.

With this process, we can obtain the nontrivial zeros with precision by maximizing the trigonometric partitions and the final angle. The positive equation of the quadratic solution of the Riemann Zeta function is one of the most consistent in tending toward the value $\mathbf{g} = \mathbf{1/2}$, which represents an asymptote of the graph.

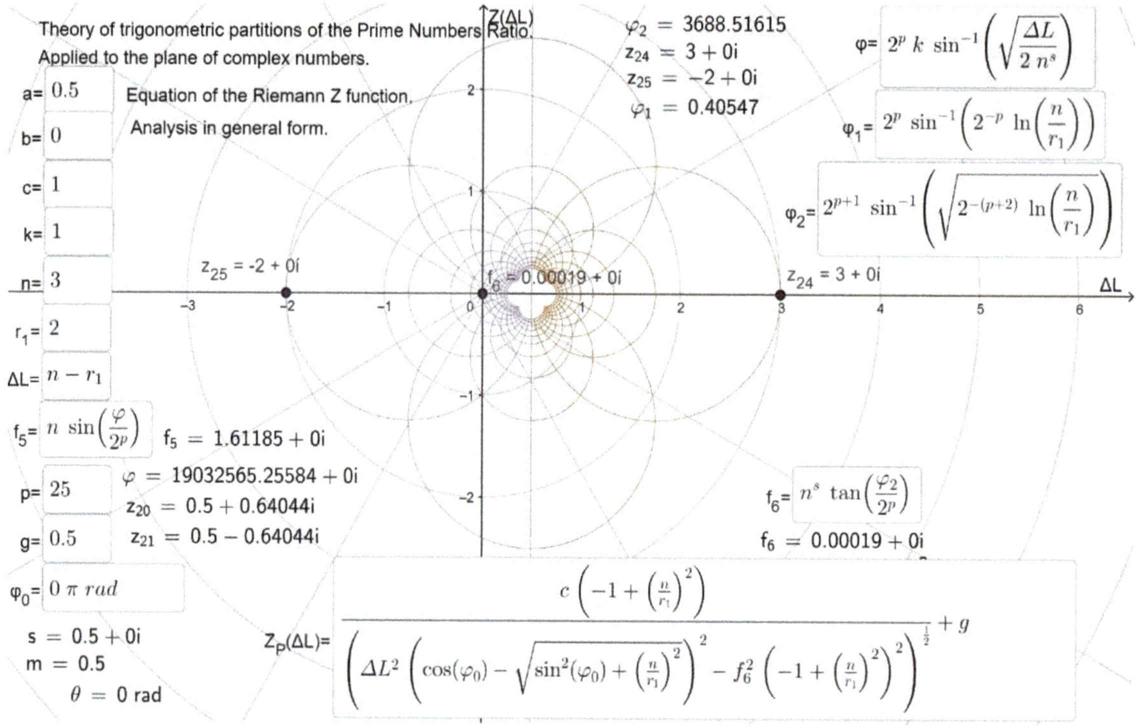

Figure 2.73: Relation Between Prime Numbers and the Riemann Zeta Function. Analysis in the Complex Number Plane for Negative Equations.

2.4 Annexes: List Of Videos Of The Theory Of Trigonometric Partitions and Riemann's Z(s) Function:

https://youtu.be/p0Zc9onKQ0Q?si=Z3-dRkEOZkbyRWy0

https://youtu.be/3Ebvypj577E?si=idU9kv2Cz7JWtvxs

https://youtu.be/pH-0CcdHP84?si=T8_k2-hIoEzg6DJt

https://youtu.be/Y_eQAANhI4Q?si=8VppT4WPcK3mYw-d

https://youtu.be/zTQHj0EhNmM?si=S216L2D1jULH3yAe

https://youtu.be/ZsHNJjjYspk?si=AesMVnThMIPMkYpz

https://youtu.be/BRBHCn-jb9w?si=lFKDzlemIAbeaY49

https://youtu.be/Xd8V9ST1RDg?si=REKYp1hVmApTISz0

https://youtu.be/E9TFr1RLmEA?si=FFqRNJ0IaVdP8b_z

https://youtu.be/obQpRzpDyRA?si=Mwdqn_Ji9oq9V8tC

https://youtu.be/Lk3n2cQpiF0?si=CBWJQT5rfweyJ8Gv

https://youtu.be/QieWM4IlNmM?si=aAhM9TbF1rlGV9ol

https://youtu.be/BBzZt-NWdqA?si=4H-BB1nEkIYM1J76

https://youtu.be/YqthYT9vypk

https://youtu.be/RlIXGEaG3NE

https://youtu.be/FZgm6vV9za4

https://youtu.be/9BxkQrD5BVs

2.5 Social Networks

YouTube:

https://www.youtube.com/@josephdays07

Linked In:

https://www.linkedin.com/in/jos%C3%A9-mauricio-orellana-d%C3%ADaz-71b671200

Instagram:

https://www.instagram.com/josemauricio.orellanadiaz?r=nametag

TikTok:

tiktok.com/@josephdays07

References

Clay Mathematics Institute. (n.d.-a). *Riemann hypothesis.* https://www.claymath.org/lectures/riemann-hypothesis/. (Accessed: 2024-05-16)

Clay Mathematics Institute. (n.d.-b). *Riemann hypothesis.* https://www.claymath.org/collections/riemanns-1859-manuscript/. (Accessed: 2024-05-16)

Edwards, H. M. (2001). *Riemann's zeta function* (Vol. 58). Courier Corporation.

Orellana Díaz, J. M. (2023). *Theory of spiral angles, spirals and trigonometric partitions* (Vol. 1). Barker and Joules.